WIRELESS TECHNOLOGY PROSPECTS AND POLICY OPTIONS

Committee on Wireless Technology Prospects and Policy Options

Computer Science and Telecommunications Board

Division on Engineering and Physical Sciences

NATIONAL RESEARCH COUNCIL
OF THE NATIONAL ACADEMIES

THE NATIONAL ACADEMIES PRESS
Washington, D.C.
www.nap.edu

THE NATIONAL ACADEMIES PRESS   500 Fifth Street, N.W.   Washington, DC 20001

NOTICE: The project that is the subject of this report was approved by the Governing Board of the National Research Council, whose members are drawn from the councils of the National Academy of Sciences, the National Academy of Engineering, and the Institute of Medicine. The members of the committee responsible for the report were chosen for their special competences and with regard for appropriate balance.

Support for this project was provided by the National Science Foundation under award number CNS-0238131. Any opinions, findings, conclusions, or recommendations expressed in this publication are those of the authors and do not necessarily reflect the views of the organization that provided support for the project.

International Standard Book Number-13: 978-0-309-16398-9
International Standard Book Number-10: 0-309-16398-6

Copies of this report are available from:

The National Academies Press
500 Fifth Street, N.W., Lockbox 285
Washington, DC 20055
(800) 624-6242
(202) 334-3313 (in the Washington metropolitan area)
Internet: http://www.nap.edu

Copyright 2011 by the National Academy of Sciences. All rights reserved.

Printed in the United States of America

THE NATIONAL ACADEMIES
Advisers to the Nation on Science, Engineering, and Medicine

The **National Academy of Sciences** is a private, nonprofit, self-perpetuating society of distinguished scholars engaged in scientific and engineering research, dedicated to the furtherance of science and technology and to their use for the general welfare. Upon the authority of the charter granted to it by the Congress in 1863, the Academy has a mandate that requires it to advise the federal government on scientific and technical matters. Dr. Ralph J. Cicerone is president of the National Academy of Sciences.

The **National Academy of Engineering** was established in 1964, under the charter of the National Academy of Sciences, as a parallel organization of outstanding engineers. It is autonomous in its administration and in the selection of its members, sharing with the National Academy of Sciences the responsibility for advising the federal government. The National Academy of Engineering also sponsors engineering programs aimed at meeting national needs, encourages education and research, and recognizes the superior achievements of engineers. Dr. Charles M. Vest is president of the National Academy of Engineering.

The **Institute of Medicine** was established in 1970 by the National Academy of Sciences to secure the services of eminent members of appropriate professions in the examination of policy matters pertaining to the health of the public. The Institute acts under the responsibility given to the National Academy of Sciences by its congressional charter to be an adviser to the federal government and, upon its own initiative, to identify issues of medical care, research, and education. Dr. Harvey V. Fineberg is president of the Institute of Medicine.

The **National Research Council** was organized by the National Academy of Sciences in 1916 to associate the broad community of science and technology with the Academy's purposes of furthering knowledge and advising the federal government. Functioning in accordance with general policies determined by the Academy, the Council has become the principal operating agency of both the National Academy of Sciences and the National Academy of Engineering in providing services to the government, the public, and the scientific and engineering communities. The Council is administered jointly by both Academies and the Institute of Medicine. Dr. Ralph J. Cicerone and Dr. Charles M. Vest are chair and vice chair, respectively, of the National Research Council.

www.national-academies.org

COMMITTEE ON WIRELESS TECHNOLOGY PROSPECTS AND POLICY OPTIONS

DAVID E. LIDDLE, U.S. Venture Partners, *Chair*
YOCHAI BENKLER, Harvard University
DAVID BORTH, Motorola Labs
ROBERT W. BRODERSEN, University of California, Berkeley
DAVID D. CLARK, Massachusetts Institute of Technology
THOMAS (TED) DARCIE, University of Victoria
DALE N. HATFIELD, University of Colorado, Boulder
MICHAEL L. KATZ, New York University
PAUL J. KOLODZY, Kolodzy Consulting
LARRY LARSON, University of California, San Diego
DAVID P. REED, Massachusetts Institute of Technology
GREGORY ROSSTON, Stanford University
DAVID SKELLERN, National ICT Australia

Staff

JON EISENBERG, Director, Computer Science and
    Telecommunications Board

COMPUTER SCIENCE AND TELECOMMUNICATIONS BOARD

ROBERT F. SPROULL, Oracle Corporation, *Chair*
PRITHVIRAJ BANERJEE, Hewlett-Packard Company
STEVEN M. BELLOVIN, Columbia University
SEYMOUR E. GOODMAN, Georgia Institute of Technology
JOHN E. KELLY III, IBM
JON M. KLEINBERG, Cornell University
ROBERT KRAUT, Carnegie Mellon University
SUSAN LANDAU, Radcliffe Institute for Advanced Study
DAVID E. LIDDLE, U.S. Venture Partners
WILLIAM H. PRESS, University of Texas, Austin
PRABHAKAR RAGHAVAN, Yahoo! Labs
DAVID E. SHAW, D.E. Shaw Research
ALFRED Z. SPECTOR, Google, Inc.
JOHN A. SWAINSON, Silver Lake
PETER SZOLOVITS, Massachusetts Institute of Technology
PETER J. WEINBERGER, Google, Inc.
ERNEST J. WILSON, University of Southern California

Staff

JON EISENBERG, Director
VIRGINIA BACON TALATI, Associate Program Officer
SHENAE BRADLEY, Senior Program Assistant
RENEE HAWKINS, Financial and Administrative Manager
HERBERT S. LIN, Chief Scientist
EMILY ANN MEYER, Program Officer
LYNETTE I. MILLETT, Senior Program Officer
ERIC WHITAKER, Senior Program Assistant
ENITA A. WILLIAMS, Associate Program Officer

For more information on CSTB, see its website at http://www.cstb.org, write to CSTB, National Research Council, 500 Fifth Street, N.W., Washington, DC 20001, call (202) 334-2605, or e-mail the CSTB at cstb@nas.edu.

Preface

The use of radio-frequency communication—commonly referred to as wireless communication—is becoming more pervasive as well as more economically and socially important. Technological progress over many decades has enabled the deployment of several successive generations of cellular telephone technology, which is now used by many billions of people worldwide; the near-universal addition of wireless local area networking to personal computers; and a proliferation of actual and proposed uses of wireless communications. The flood of new technologies, applications, and markets has also opened up opportunities for examining and adjusting the policy framework that currently governs the management and use of the spectrum and the institutions involved in it, and models for allocating spectrum and charging for it have come under increasing scrutiny.

Yet even as many agree that further change to the policy framework is needed, there is debate about precisely how the overall framework should be changed, what trajectory its evolution should follow, and how dramatic or rapid the change should be. Many groups have opinions, positions, demands, and desires related to these questions—reflecting multiple commercial, social, and political agendas and a mix of technical, economic, and social perspectives.

The development of technologies and associated policy and regulatory regimes are often closely coupled, an interplay apparent as early as the 1910s, when spectrum policy emerged in response to the growth of radio communications. As outlined in this report, current and ongoing technological advances suggest the need for a careful reassessment of the assumptions that inform spectrum policy in the United States today.

This report of the Committee on Wireless Technology Trends and Policy Options (Appendix A) thus seeks to shine a spotlight on 21st-century technology trends and to outline the implications of emerging technologies for spectrum management in ways that the committee hopes will be useful to those setting future spectrum policy. Speakers at the meetings held by the committee are listed in Appendix B. The detailed statement of task for the study is given in Appendix C.

The committee was not in a position to examine details of the numerous specific areas of contention that are the subject of frequent debate today or to evaluate the merits of opposing claims. This report thus does not offer specific prescriptions for how particular frequency bands should be used or seek to resolve conflicting demands for spectrum use for particular services. Instead, the committee offers a discussion of the technology trends and related policy options relevant to addressing these conflicts, both today and in the future.

The development of this report was not without its own challenges, and the report was a long time in the making. Early on, the committee's work expanded in scope following a request from the National Telecommunications and Information Administration to convene a forum on spectrum policy reform options.[1] Later, a variety of circumstances unrelated to the substance or the work of the committee led to unexpected delays. Throughout the project, there were also reminders that its subject is inherently complex and challenging. The technology and policy issues are tightly intertwined, and the study involved experts from multiple disciplines, including economics, law, public policy, electrical engineering, and computer science. The multidisciplinary approach sought yields a more comprehensive view of a problem, but more time and effort are needed to establish a common view of the issues, a common vocabulary, and so forth. Finally, the technical and policy perspectives of the members of the committee were, by design, diverse. As a result, the technology considerations, enablers of a more nimble policy framework, and policy options developed by the committee are the products of a multidimensional examination of the issues and negotiation of agreements among members holding often-contrasting opinions.

---

[1] National Research Council, *Summary of a Forum on Spectrum Management Policy Reform*, The National Academies Press, Washington, D.C., 2004.

Acknowledgment of Reviewers

This report has been reviewed in draft form by individuals chosen for their diverse perspectives and technical expertise, in accordance with procedures approved by the National Research Council's Report Review Committee. The purpose of this independent review is to provide candid and critical comments that will assist the institution in making its published report as sound as possible and to ensure that the report meets institutional standards for objectivity, evidence, and responsiveness to the study charge. The review comments and draft manuscript remain confidential to protect the integrity of the deliberative process. We wish to thank the following individuals for their review of this report:

Vinton G. Cerf, Google, Inc.,
John M. Cioffi, Stanford University,
Gerald R. Faulhaber, University of Pennsylvania,
Kevin C. Kahn, Intel Corporation,
Teresa H. Meng, Stanford University,
Dipankar Raychaudhuri, Rutgers University,
David H. Staelin, Massachusetts Institute of Technology,
Andrew J. Viterbi, The Viterbi Group, and
Steven S. Wildman, Michigan State University.

Although the reviewers listed above have provided many constructive comments and suggestions, they were not asked to endorse the conclusions or recommendations, nor did they see the final draft of the report before its release. The review of this report was overseen by R. Stephen Berry, University of Chicago. Appointed by the National Research Council, he was responsible for making certain that an independent examination of this report was carried out in accordance with institutional procedures and that all review comments were carefully considered. Responsibility for the final content of this report rests entirely with the authoring committee and the institution.

Contents

SUMMARY   1

1   INTRODUCTION: TRENDS AND FORCES RESHAPING   14
    THE WIRELESS WORLD
    Advances in Radio Technology, 15
    Expansion in Applications and Users, 17
    Changing Market Dynamics, 20
    The Evolving Policy and Regulatory Framework, 21

2   KEY TECHNOLOGY CONSIDERATIONS   33
    Technological Advances in Radios and Systems of Radios, 34
    Low-Cost, Portable Radios at Frequencies of 60 GHz and
        Above, 52
    Interference as a Property of Radios and Radio Systems,
        Not Radio Signals, 53
    Enduring Technical Challenges, 55
    Timescales for Technology Deployment, 57
    Talent and Technology Base for Developing Future Radio
        Technology, 58
    Measurements of Spectrum Use, 59
    Challenges Facing Regulators, 63
    Engineering Alone Is Often No Solution, 66

3   POLICY OPTIONS   67
     Pressures on Today's Wireless Policy Framework, 67
     Key Considerations for a Future Policy Framework, 68
     Technology-Enabled Policy Options, 76

APPENDIXES

A   Biographies of Committee Members and Staff   87
B   Speakers at Meetings   96
C   Statement of Task   99

Summary

Today's framework for wireless policy—which governs the operation of devices that make use of radio-frequency (RF) transmissions—has its roots in the technology of 80 years ago and the desire at that time for governmental control over communications. It has evolved to encompass a patchwork of legacy rules and more modern approaches that have been added over time. Although views vary considerably on whether the pace of reform has been commensurate with the need or opportunity, there have been a number of significant policy changes in recent decades to adjust to new technologies and to decrease reliance on centralized management. These developments have included the use of auctions to make initial assignments (along with the creation of secondary markets to trade assignment rights) and the designation of open bands[1] in which all users are free to operate subject only to a set of "rules of the road."

There remains, nonetheless, much debate about how the overall framework should be changed, what trajectory its evolution should follow, and how dramatic or rapid the change should be. Many groups have opinions, positions, and demands related to these questions reflecting multiple commercial, social, and political agendas and a mix of technical, economic, and social perspectives.

---

[1] A variety of terms are used to describe this approach, including "license-exempt" or "license by rule." The approach is probably most familiar as the basis for operation of wireless LANs, cordless telephones, and the like.

PRESSURES ON TODAY'S WIRELESS POLICY FRAMEWORK

The current framework for wireless policy in the United States is under pressure on several fronts:

- It continues to rely heavily on service-specific allocations and assignments that are made primarily by frequency band and geographic location and does not embrace all of the spectrum management approaches possible with today's technologies and expected to be available with tomorrow's technologies.
- Despite revisions aimed at creating greater flexibility, it continues to rely significantly on centrally managed allocation and assignment, with government regulators deciding how and by whom wireless communications are to be used despite growing agreement that central management by regulators is inefficient and insufficiently flexible.
- It will not be able to satisfy the increasing and broadening demand for wireless communications that is spurred by interest in richer media, seemingly insatiable demand for mobile and untethered access to the Internet and the public telephone network, and growing communication among devices as well as people.
- It does not fully reflect changes in how radios are being built and deployed now or in how they could be built and deployed in the future in response to different regulations, given that technological innovation has expanded the range of potential wireless services and the range of technical means for providing those services and at the same time has dramatically lowered the cost of including wireless functionality in devices.

Today, the complexity and density of existing allocations, assignments, and uses, and the competing demands for new uses, all make policy change difficult. Decisions will necessarily involve (1) addressing the costs and benefits of proposed changes that are (often unevenly) distributed over multiple parties, (2) resolving conflicting claims about costs and benefits, and (3) addressing coordination issues, which are especially challenging if achieving a particular change requires actions by a large number of parties. Moreover, some parties stand to gain by changing—or advocating for change—while others stand to gain by delay or retaining the status quo.

FORWARD-LOOKING POLICY DIRECTIONS

The Committee on Wireless Technology Prospects and Policy Options believes that, moving forward, the unambiguous goal for spectrum policy should be to make the effective supply of spectrum plentiful so as to make it cheaper and easier to innovate and introduce new or enhanced services. Put another way, the goal should be to reduce the total cost—which includes the cost, if any, of licenses, and the cost of equipment, both for the end user and the network—of introducing or enhancing services. The financial cost of adverse impacts to existing users and services should also be fairly evaluated and debated in advance of regulatory changes.

Given the plethora of existing allocations and assignments, and the multitude of existing services and users associated with them, it is not possible to take a clean-slate approach. Achieving the goal stated above will thus involve several parallel efforts:

- *Leveraging advanced technologies, regulation, and market-based incentives to support sharing,* including overlay and underlay approaches, so that new services can share spectrum with legacy services.
- *Streamlining and modernizing the use of bands allocated or assigned to old services to free up new areas of "white space" that can be used for new services,* by using market mechanisms, relinquishing government-controlled bands used for obsolete services, and shutting down obsolete services (as has happened with analog television).
- *Establishing "open" as the default policy regime used at 20 to 100 gigahertz (GHz).* At these higher frequencies, sparser use and technical characteristics that significantly reduce the chance for interference suggest that nontraditional management approaches can predominate.

The likelihood of ongoing technological change also points to the value of establishing a more adaptive learning system for setting policy that would be better able to track and even anticipate advances in wireless technology and emerging ways of implementing and using wireless services.

The sections that follow provide a brief description of key technology considerations and outline policy options, many enabled by new technology, that will be useful in achieving the goal of increasing the supply of spectrum for enhanced or new services.

KEY TECHNOLOGY CONSIDERATIONS

Radio-frequency communication has been transformed profoundly in recent years by a number of technological advances. This section outlines key recent advances and associated trends and their implications for the design of radios and radio systems and for regulation and policy.

Profound Changes in Radio-Frequency Communication as a Result of Technological Advances in Radios and Radio Systems

Digital processing is used increasingly to detect the desired signal and to reject interfering signals. The shift to largely digital radios built using complementary metal oxide semiconductor (CMOS) technology; (a high-density, low-power-consumption technology for constructing integrated circuits) has made it much cheaper and easier to include wireless capabilities in consumer electronic devices. As a result of the reduction in costs for radio technology, *the barriers to developing and deploying novel, low-cost, specialized radios have become much lower, and more firms and other organizations have become capable of and potentially motivated to participate.* Growth in the number of wireless devices of various types and in the demand for wireless communications is likely to continue.

Technological capabilities are also driving the introduction of new radio system architectures, including a *shift away from centralized systems to more localized transmissions in distributed systems* that use very small cells (the smallest of those being deployed today are called femtocells) or mesh networks, and a shift from centralized switching to more distributed, often Internet-Protocol-based, networks.

Another important shift in radios has been the ability to use new techniques to permit *greater dynamic exploitation of all available degrees of freedom—frequency, space, time, and polarization*—which makes it possible to take greater advantage in a dynamic, fine-grained, and automated fashion of all the degrees of freedom to distinguish signals. This capability offers the opportunity to introduce new options for assigning usage rights.

The ability to leverage sustained improvements in the performance of digital logic also opens up opportunities to build *radios that are much more flexible and adaptable.* Such radios can change their operating frequency and modulation scheme, can sense and respond to their environment, and can operate cooperatively to create new opportunities to make more dynamic, shared, and independently coordinated use of spectrum. (They cannot, however, directly sense passive users, which means that special measures such as registries or beacons are needed for detection of passive users.) The result is that radios and systems of radios can operate and cooperate in an increasingly dynamic and autonomous manner.

Although increased flexibility involves greater complexity, cost, and power consumption, it *enables building radios that can better coexist with existing radio systems,* through both underlay (low-power use intended to have a minimal impact on the primary user) and overlay (agile use by a secondary user of "holes" in the time and space of use by the primary user). Moreover, flexibility makes it possible to *build radios with operating parameters that can be modified to comply with future policy or rule changes or future service requirements.*

The use of CMOS to build radios and digital processing together with other advances in RF technology opens up a new set of opportunities in the form of *low-cost, portable radios that are becoming increasingly practical at frequencies of 60 GHz and above.* Radios operating in this domain must confront a number of challenges, including limited free-space propagation distances (especially in the oxygen absorption bands around 60 GHz) and very limited penetration through and diffraction around walls of buildings or other obstacles. On the other hand, these characteristics make such radios very useful in providing very large bandwidths over short range.

Interference as a Property of Radio Receivers and Radio Systems, Not Radio Signals

It is commonplace to talk about radio signals interfering with each other, a usage that mirrors the common experience of hearing broadcast radio signals that are transmitted on the same channel overlay each another. However, radio signals themselves do not, generally speaking, interfere with each other in the sense that information is destroyed. Interference reflects a receiver's inability to distinguish between the desired and undesired signals. *The cost of separating these signals is ultimately reflected in design complexity, hardware cost, and power consumption.* As a result, any practical radio (i.e., one of practical size, cost, and power consumption) will necessarily throw away some of the information needed to resolve signal ambiguity. As the performance and capabilities of radios continue to improve over time, their ability to distinguish between signals can be expected to improve. However, power consumption will remain an especially challenging constraint, especially for portable devices, and even a modest additional device cost can jeopardize the commercial viability of a product or service.

Persisting Technical Challenges

Even as the capabilities and the performance of radios continue to improve, several hard technical problems can be expected to persist. These technical challenges—discussed in more detail below in this report—include power consumption, nonlinearity of radio components, support for nomadic operation and mobility, and coping with the heterogeneity of capabilities, including both legacy equipment and systems that are inherently constrained, such as embedded network sensors and scientific instruments that passively use spectrum (e.g., for remote Earth sensing and radio astronomy).

Nonuniform Timescales for Technology Replacement

Different wireless services are characterized by the different timescales for removal of old technology from service and deployment of new technology. The factors influencing the turnover time include the time to build out the infrastructure, the time to turn over the base of end-user devices, and the time to convince existing users (who may be entrenched and politically powerful) to make—and pay for—a shift, as well as the incentives for upgrading and the size of the installed base.

Considerable Uncertainty About the Rate at Which New Technologies Can Be Deployed Practically

A particular challenge in contemplating changes to policy or regulatory practice is determining just how quickly promising new technologies will be deployable as practical devices and systems and thus how quickly, and in what directions, policy should be adjusted. As is natural with all rapidly advancing technologies, the *concepts and prototypes are often well ahead of what has been proved to be technically feasible or commercially viable.* At the same time, technical advances sometimes can be commercialized quickly, although deployment and use might also require adjustments to regulations, a process that historically has taken longer.

Spectrum Use Lower Than Allocations and Assignments Suggest, Especially at Higher Frequencies

Quantifying how well and how efficiently spectrum is used is quite challenging. Measurements may miss highly directional or periodic use and cannot detect passive uses such as radio astronomy. These caveats notwithstanding, measurements suggest that some allocated and assigned frequency bands are very heavily used whereas others are only lightly used, at least in certain places and at certain times. The published frequency allocation and assignment charts are thus potentially misleading in their suggestion that little spectrum is available for new applications and services. A good deal of empty space exists in the spectrum; the challenge is to find ways of safely detecting and using it.

ENABLERS OF A MORE NIMBLE, FORWARD-LOOKING SPECTRUM POLICY FRAMEWORK

The committee identified the following approaches as enablers of a more nimble approach to spectrum policy.

Abandon the Extremes in the "Property Rights" Versus "Commons" Debate

The terms "property rights" and "commons" are shorthand for particular approaches to spectrum management—approaches that reflect philosophical and ideological perspectives as well as technical and policy alternatives. The property rights approach relies on a well-specified and possibly exclusive license to operate and on rights that can be established or transferred through an administrative proceeding, auction, or market transaction. It is intended to facilitate the creation of a market in infrastructure access and use rights. The commons or open-access approach relies on establishing license-free bands in which users must comply with specified rules, such as limits on transmitted power. It is intended to facilitate a market in devices and services based on symmetrically applied infrastructure use and access rights.

Each has advantages and disadvantages and associated transaction costs. Each involves different incentives, and different and complementary loci, for innovation. When carefully specified, neither pure version can at present be determined to be uniquely "better" than the other. Moreover, there is a much larger space of alternatives, and commercial forces can help drive their evolution and selection provided that the regulatory structure is not overly rigid. This suggests adopting a policy framework that avoids detailed allocation of spectrum in favor of one that uses market mechanisms for spectrum allocation where they make sense and uses an open-access mechanism in other instances. Where to draw the line between the two general approaches (licensed or exclusive-use allocations versus open access)—and which hybrids of the two approaches might be useful—will shift as technological capabilities, deployed services, and business models continue to evolve.

Leverage Standards Processes but Understand Their Limitations

Regulators often rely, either explicitly or implicitly, on standards bodies to define the technical standards that are ultimately needed to implement rulings for proposed new allocations and services. On the one hand, standards-setting organizations are viewed as being more nimble and better able than regulatory bodies to focus on technical issues. On the other hand, as standards take on greater importance, the number of competing players and conflicting interests grows, raising the risks that a large player may try to dominate the process, that standards setting may deadlock, or that only certain societal interests are reflected. Some ways to address these risks have been identified, such as the use of one company, one vote to deal with attempts to dominate by sending multiple delegates, but such an approach has tradeoffs as well.

Collect More Data on Spectrum Use

There are many gaps today in knowledge about the use of spectrum. Measuring use is difficult and has not been done systematically, leading to uncertainty for policy makers, who are not able to readily assess claims and counterclaims about the use or nonuse of spectrum. Advances in radio technology, however, make it possible to contemplate new ways of collecting data on spectrum use, such as by the deployment of networks of sensors and the incorporation of sensing capabilities in equipment deployed for other purposes. Such capabilities would enhance the ability of regulators to enforce compliance with operating rules, and to more quickly assess conflicting claims about harmful interference and provide the data required to implement spectrum management schemes that depend on identifying unused spectrum.

Ensure That Regulators Have Access to Technology Expertise Needed to Address Highly Technical Issues

As this report argues, spectrum policy is entering an era in which technical issues are likely to arise on a sustained basis as technologies, applications, and services continue to evolve. The committee believes that the Federal Communications Commission (FCC) would therefore benefit from enhancing its technology assessment and engineering capabilities and suggests several ways to gain such expertise:

- Make it a priority to recruit top-caliber engineers/scientists to work at the FCC, perhaps for limited terms.
- Use an external advisory committee to provide the FCC with outside, high-level views of key technical issues. (Indeed, in the past, the FCC convened the Technology Advisory Council to play just such a role.[2])
- Add technical experts to the staff of each commissioner.
- Tap outside technical expertise, including expertise elsewhere in the federal government such as at the Department of Commerce's Institute for Telecommunication Sciences and the National Institute of Standards and Technology (NIST), or through a federally funded research and development center.

---

[2] The FCC announced the appointment of a new Technology Advisory Council in October 2010, as this report was being prepared for publication.

Sustain Talent and Technology Base for Future Radio Technology

The opportunities described in this report rely on innovation in both technology and policy. Innovation in wireless technology involves many areas of science and engineering—including RF engineering, digital logic, CMOS, networking, computer architecture, applications, policy, and economics—and often expertise in combinations of these areas that is difficult to obtain in a conventional degree program. Research investments in wireless technologies by federal agencies such as the National Science Foundation, Defense Advanced Research Projects Agency, National Telecommunications and Information Administration, and NIST help to build the knowledge base for future innovation and to educate and train tomorrow's wireless engineering talent. Research efforts can be buttressed by an infrastructure for implementing and testing new ideas in radios and systems of radios. Test beds allow radio system architectures to be tested at scale, and access to facilities for integrated circuit design and fabrication makes it possible to build prototypes.

FORWARD-LOOKING POLICY OPTIONS

Consider "Open" as the Default Policy Regime at a Frequency Range of Approximately 20 to 100 GHz

At frequencies of 20 to 100 GHz, the potential for legacy problems and for interference (in the classical sense) is lower, suggesting that nontraditional (open) approaches could predominate for use of spectrum at 20 to 100 GHz.[3] Adopting an open approach for a frequency domain that will become increasingly more technologically accessible and commercially attractive several years from now would set the stage for more flexible and adaptive future spectrum management. FCC policy has already moved in this general direction, with an unlicensed regime established in a band at 57 to 64 GHz and licensed access to bands at 80 and 95 GHz made available on a first-come, first-protected basis.

Spectrum use is relatively low at 20 to 100 GHz compared to use at frequencies below 20 GHz, but existing users are likely to argue vociferously for ongoing protection, and some exceptions to the open rule will probably be needed to protect certain established services and passive scientific uses.

---

[3] It would be imprudent to recommend a particular regime for frequencies above 100 GHz given today's limited understanding of how radios might be constructed or operated in that domain, and it would be prudent to review policy in this area every several years and make adjustments as appropriate.

Use New Approaches to Mitigate Interference and a Wider Set of Parameters in Making Assignments

Protecting against harm from interference has both technical aspects (how well a radio or radio system can separate the desired from undesired signals) and economic dimensions (the costs of building, deploying, and operating a radio or radio system with particular technical characteristics that make it easier to separate the signals).

Provided that the transaction costs are low enough and that agreed-upon protocols for coordination exist, usage "neighbors" can negotiate mutually satisfactory solutions to interference problems that take into account the financial benefits, costs, and technology opportunities.[4] Given the complexity of defining the technological options for any given communication in the context of other local attempts to communicate, as well the difficulties of determining who is a "neighbor," particularly for mobile and nomadic systems, the transaction costs may be significant.[5] The size of these costs and their implications for solutions that rely on negotiations will depend on such factors as the number and diversity of systems and users and is a subject of ongoing debate.

Receivers are increasingly able to discriminate a desired signal from an undesired one, some technologies provide new tools for mitigating interference, and other new technologies make it possible to exploit all degrees of freedom in a dynamic fashion, opening new avenues for mitigating interference. Mitigation of interference can also be addressed in terms of the behavior of systems of radios rather than of individual radios and by coordinating the behavior of multiple systems. A key question is how best to establish incentives for such cooperation.

Introduce Technological Capabilities That Enable More Sophisticated Spectrum Management

The use of certain technologies, some of them emerging and some of them available but not widely deployed, would make it easier to introduce new services into crowded frequency bands. In particular it might be possible to overlay unlicensed use onto licensed use if receivers were suitably equipped. Another enabling technology is smart antennas that could be used to focus transmitted power, scan the environment for other transmissions, and spatially separate transmissions to help avoid interference. Migrating current nondigital services to more efficient digital

---

[4] R.H. Coase, "The Federal Communications Commission," *Journal of Law and Economics* 2(10):1-40, 1959.

[5] Y. Benkler, "Some Economics of Wireless Communications," *Harvard Journal of Law and Technology* 16(1):25-83, 2002.

transmission will be a major challenge, especially for services that have large and/or politically powerful legacy bases.

Migrating to higher-quality receivers has a cost in dollars, design complexity, and power consumption. Even small additional costs matter a great deal when service providers are fighting for pennies. But the additional investment could have a big payoff for those who seek to introduce enhanced or new services.

Trade Near-Absolute Outcomes for Statistically Acceptable Outcomes

Although statistical models have long been used in spectrum analysis, the underlying conservative assumptions have emphasized avoidance of interference to an extent that has significantly affected efficient use of spectrum. An alternative is to relax constraints so as to normally (but not always) provide good outcomes, as is done in both Internet communication (best-effort packet delivery) and cellular telephony (which provides mobility in exchange for gaps in coverage and lower audio quality). With this approach, adverse impacts on users would be rare even though technical performance might be measurably but tolerably worse for users. A relaxation of requirements could significantly open up opportunities for nonexclusive use of frequency bands through a rebalancing of the risk of interference and the benefits of new services. This approach might not be appropriate, however, for services that demand guarantees of especially high-quality service (e.g., for certain safety-critical systems). Although regulatory proceedings could be used to implement such a shift, it might be preferable for licensees to negotiate mutually beneficial arrangements.

Design for Light as Well as Design for Darkness

Many systems, notably cellular phones, have been "designed for darkness"—that is, with the assumption that a particular band has been set aside for a particular service or operator and that there are no other emissions in that band. An alternative is to "design for light," with the assumption that the operating environment will be noisy and cluttered. Both approaches are reasonable for certain applications and services, but there are tradeoffs between (1) the ease with which higher spectral efficiency can be achieved under design for darkness, thus allowing for lower cost and reduced power consumption and (2) the greater flexibility to support multiple and diverse uses under design for light. The historical preference has been to design for darkness, but today technological advances suggest opening up more bands in the design-for-light modality.

Consider Regulation of Receivers and Networks of Transceivers

Much regulation has focused on transmitters, and rules have specified transmission frequency and bandwidth, geographical location, and transmission power. Increasing use of new radio architectures (discussed above) suggests that the scope of inquiry can be broadened to look at the properties and behaviors of receivers and networks of transceivers. Better receiver standards would create an environment in which receiver capabilities present a lower barrier than they do today for implementing new spectrum-sharing schemes. Expanding the scope of policy or regulation to include a system of radios rather than an individual radio would open up new opportunities, such as the possibility of exploiting a network of radios to reliably use a listen-before-send protocol to avoid interference and thereby avoid the hidden node problem, in which one radio cannot detect transmissions from another radio.

Exploit Programmability So That Radio Behavior Can Be Modified to Comply with Operating Rule Changes

Because radios can be made highly programmable, albeit with tradeoffs in complexity, cost, and power consumption, their operating parameters can be made modifiable to comply with policy or rule changes. Deployment of devices with such capabilities opens up new opportunities for more flexible regulation and more incremental policy making: (1) policies could be written less precisely up front, (2) policies would not have to be homogeneous and could be adapted to local environmental conditions such as signal density, (3) the operating rules of existing devices could be revised to accommodate new technology, and (4) devices could more easily be certified for international use because they can readily be switched to comply with local policy. One result could be greater speed of deployment for new technologies and services.[6] Over time, the introduction of such capabilities could be expected to impose a less onerous performance and cost penalty. Future regulations could take advantage of this opportunity by specifying, for example, that licenses granted after a certain date would require use of devices with a certain degree of reprogrammability.

---

[6] Caveat: this flexibility could also paradoxically represent a disincentive to deployment because it opens up the possibility of future forced sharing, potentially reducing the value of a particular license.

Use Adaptive and Environment-Sensing Capabilities to Reduce the Need for Centralized Management

As agility, sensing, and coordination capabilities improve and as etiquettes and standards for these capabilities develop, opportunities will arise for scaling back centralized management. Potential advantages of this approach include a lower barrier to entry (because neither engagement with a regulator for spectrum assignment nor negotiation with an existing license holder would be necessary) and greater flexibility of use (because operation would be defined primarily by the attributes of radio equipment rather than regulation). Potential disadvantages of this approach include uncertainty about the technical feasibility and the costs of building more capable radios with the degree of agility, coordination, and environmental sensing required for effective decentralized operation. Such a shift would also involve assessing tradeoffs between the more rapid introduction of services made possible in a decentralized regime and the significant capital investment made and efficiencies achieved, at least in some instances, under a centralized regime.

Establish Enhanced Mechanisms for Dealing with Legacy Systems

In recent years, notable efforts to deal with legacy systems have included relocating point-to-point microwave services to allow deployment of personal communications service cellular telephony and the relocation of Nextel cell services out of public safety bands. More recently, relocation of government services as well as broadcast radio services and fixed services has been undertaken to allow the introduction of new 3G/advanced wireless services bands. Modifying infrastructure to accommodate such change can be difficult and expensive; an even bigger legacy challenge is the need to migrate potentially millions of devices owned and operated by consumers and other end users. This task has proven easier when the market dynamics are such that end-user technology is regularly refreshed (as in mobile telephony, where new handsets with new features enter the market frequently and where the cost of handsets is often partly covered in the services fees and regular upgrades are made available at little additional cost to the subscriber) and harder where retrofitting is not practical and hardware has historically had a long lifetime (as in aircraft and public safety radios). The difficulty of making changes also depends, of course, on the relative political clout of the incumbents and those seeking to introduce new services.

1

Introduction: Trends and Forces Reshaping the Wireless World

This report examines the evolution of radio-frequency communication—commonly referred to as wireless communication[1]—and the framework that governs its use (a framework that also extends to uses of radio frequencies for purposes other than communication). An avalanche of new technologies, applications, and markets for wireless communications is colliding with a well-established and comprehensive but increasingly obsolescent framework for the allocation, assignment, and utilization of the radio spectrum. Even as demand for wireless services continues to grow, much of the radio spectrum has already been allocated and assigned by frequency band (and often by geographical location) for a multitude of private-sector and government uses. The more recent developments come on the heels of many decades of technological progress, notably marked by widespread deployment of existing wireless capabilities such as several successive generations of cellular telephone technology now used by billions of people worldwide and a proliferation of actual and proposed uses of wireless communications.

Significant policy changes in recent decades reflect efforts to adjust to new technologies and to decrease reliance on centralized management. There is debate about how the overall framework should be changed, what trajectory its evolution should follow, and how dramatic or rapid the change should be. Many groups have opinions, positions, and demands related to these questions, reflecting multiple commercial,

---

[1] This report uses the terms "radio" and "wireless device" synonymously.

social, and political agendas and a mix of technical, economic, and social perspectives.

This report thus seeks to shine a spotlight, in ways the committee hopes will be useful to those setting future spectrum policy, on emerging technology trends and to outline policy directions that align with those trends. It aims to provide a cogent discussion of the overall rationale for changing policy, the opportunities afforded by new technologies for spectrum management, and some long-term directions for improvement in policy.

The Committee on Wireless Technology Trends and Policy Options was not in a position to examine the details of the numerous specific areas of contention that are the subject of frequent debate today regarding use of the spectrum, or to evaluate the merits of opposing claims. This report thus does not offer specific prescriptions for how particular frequency bands should be used or seek to resolve conflicting demands for spectrum for particular services. Instead, the committee intends that its discussion of the relevant technology trends and policy options should be helpful in addressing these conflicts, both today and in the future.

ADVANCES IN RADIO TECHNOLOGY

The development of technologies and the associated policy and regulatory regimes that govern their use are often closely coupled. For example, from the late 19th century until recently, the roadways for communication and transmission of information (e.g., the telephone system, broadcast television, and radio) were, like those for transporting people and physical goods, owned, managed, and regulated by a relatively small number of institutions. The concerns and assumptions underlying policies were grounded in the technical realities and economic and political imperatives of the time. The interplay between technology and policy was apparent as early as the 1910s. The growth of radio communications and the spectrum policy that emerged reflected a compromise on a framework for spectrum management.

When spectrum regulation began with the Radio Acts of 1912 and 1927 and the Communications Act of 1934, the primary obstacle to signal reception was noise. Because of the quality of components available at that time and the nature of the most popular frequency bands of the day (which were selected for their longer propagation distances), noise was a significant problem, and interference (i.e., human-generated noise from other transmissions) from other sources was regarded as intolerable and something to be avoided. Accordingly, a regulatory structure was set up that allocated frequencies with specific power levels and bandwidth masks uniquely to single broadcasters or services in a given geographic area. For the most part, the environment consisted of a small number of high-power transmitters separated by frequency and geography, and a very large number of mute receivers. Licenses granted the right to broadcast using a few kilohertz of spectrum and also provided an "address" (in the form of, for example, AM radio channel numbers) in addition to a means to avoid interference.

Today, radios routinely operate in frequency ranges where background noise is limited and dealt with rather easily. The very large number of active transceivers means that the primary challenge is separating the desired signal from the signals of all the other potentially interfering transmitters, not avoiding noise. The huge number of devices associated with many modern services means that frequencies must be shared (and that the particular frequencies in use at any given time are not apparent to the user). For example, many cell phones share a particular block of spectrum at any given time, with the sharing enabled by separation by code (code division multiple access) or time slice (time division multiple access) as well as location (which cell the phone is currently in). These challenges were not fully anticipated by traditional spectrum allocation and licensing schemes.

Moreover, in the past 50 years, a number of changes—including a fundamental new understanding of physics and information theory; vast increases in the computation that can be performed by a compact, cheap, low-power device; and improvements in analog components—have allowed for very inexpensive processing of signals in ways not contemplated when many spectrum polices were established and allocations were made.

In short, radio-frequency communication today is being profoundly changed by a related set of technological advances—both in the capabilities and performance of individual radios and in the design of networks and systems of radios. These advances, which are discussed in more detail in Chapter 2, include the following:

• A shift in favor of digital signal processing and use of low-cost complementary metal-oxide-semiconductors integrated circuit technology for both digital and analog radio components;
• The advent of new radio systems architectures that rely on distributed (and often Internet-Protocol-based) control and on more localized transmission using microcells and mesh networks, rather than traditional architectures that rely on centralized switching or wide area transmission;
• The development of a variety of techniques, including more robust receivers, antenna arrays, frequency agility, and new modulation techniques and coding algorithms, to permit dynamic, fine-grained, and

automated exploitation of all available degrees of freedom—that is, not just static separation in frequency and space but also dynamic use of frequency, time, space, and polarization—along with "code"[2]—to distinguish radio signals; and

- The development of technologies that permit flexible and adaptable radios that can sense and respond to their operating environment and can coordinate their operation in an increasingly dynamic, distributed, and autonomous fashion.

The technological advances outlined above and discussed in more detail in the next chapter call for a careful reassessment of the assumptions that underlie spectrum policy.

EXPANSION IN APPLICATIONS AND USERS

The transition from wired and fixed place-to-place communications to wireless mobile person-to-person (and device-to-device) communications has been under way for decades.[3] Radio, once confined to largely unidirectional transmissions from a small number of broadcasters to a large number of passive receivers, has blossomed to include bidirectional communication among a much larger numbers of devices.

The number of people actively using wireless communications has grown dramatically: only a couple of decades ago, there were thousands of radio and television broadcasters, a half million amateur radio operators, and a few million mobile radio users worldwide; today there are billions of mobile telephone users, hundreds of millions of wireless local area network (WLAN) users, and similarly large numbers of low-power in-home and personal devices. Many other services and products ranging from satellite television to global positioning systems (used, for instance, in automobile navigation systems) to public safety communications make use of spectrum licensed to specific companies, government agencies, or other entities.

Perhaps most familiar and notable is that there are nearly 300 million cell phone subscribers in the United States[4] and 5 billion subscribers world-

---

[2] Although it is strictly speaking a technique for exploiting the other degrees of freedom, modulation or code is often referred to as another degree of freedom because it can be used to allow separation of signals that appear to be at the same frequency, time, and space.

[3] Donald C. Cox, "Wireless personal communications: What is it?" *IEEE Personal Communications*, April 1995, pp. 20-35. This paper notes the transition occurring already as far back as 1995 due to wireless communications.

[4] "CTIA—The Wireless Association, Wireless Quick Facts: Mid-Year Figures," available at http://www.ctia.org/media/industry_info/index.cfm/AID/10323.

wide.[5] Many everyday products that have been sold by the hundreds of millions—such as cordless phones, baby monitors, security systems, garage door openers, keyless entry for automobiles, and a wide variety of WLAN products—make use of so-called open bands for which individual licenses are not required and only low-power transmissions are permitted.

These two familiar examples are notable both for their success and for their distinct features. WLAN technology enabled the rapid and flexible deployment of a wide variety of devices. Cell phones became nearly ubiquitous as a result of large capital investments and the spectral efficiency achieved by their technology. The success of the cell phone industry was predicated on the solution of an extremely difficult (indeed nearly insurmountable) engineering problem in the presence of a huge, visible, obvious, well-understood market opportunity—universal mobile telephony. In contrast, WLANs involved solving a simpler engineering problem for a market with considerable potential but less certain value.

Many wireless devices use multiple wireless systems and technologies. Cell phones now often include Bluetooth capability,[6] allowing them to connect to wireless headsets and vehicle audio systems[7] as well as the cellular telephone system. Laptop computers today may contain wireless LAN, Bluetooth, and cellular communications capabilities. A digital video recorder might connect to a home wireless network to allow sharing photographs and music from other computers on the network while also receiving broadcast signals over the air and commercial satellite television signals. Both wireless LAN and cellular capabilities are being built into new types of consumer electronics such as electronic book readers.

Military applications of wireless technology have expanded well beyond voice communications and radar systems, and many applications initially developed for military purposes have found widespread commercial or civilian use. For instance, the Global Positioning System (GPS) was launched as a military application and is now used by hikers, in-vehicle navigation systems, and even in golf carts.

More recently, wireless technology has been applied to machine-to-machine communications, with expectations that such communications will exceed those involving humans within the next few years.[8] Fleet

---

[5] Estimates were that by the end of 2010, there would be 5.3 billion mobile subscriptions worldwide. See International Telecommunication Union (ITU), *The World in 2010: ICT Facts and Figures*. Geneva.

[6] Bluetooth wireless technology is one of several short-range communications technologies intended to replace the cables connecting portable and fixed devices.

[7] The increasing prevalence of laws requiring hands-free operation of cellular phones in automobiles in the interest of safety concerns is driving increased interest in this application of wireless technology.

[8] "A World of Connections: A Special Report on Telecoms," p. 5 in *The Economist*, April 28, 2007.

management, supply chain and logistics management, automated meter reading, security monitoring systems, vending machines, and sensor networks monitoring industrial process are just a few examples of the applications already in use and being developed. These distributed control systems made up of sensors, remote devices, and actuators are linked into wireless networks via wireless communications channels.[9] Radio frequency identification (RFID) uses wireless communication to identify tagged objects. Although this prospect has been anticipated for some time,[10] such applications are now being more widely adopted. Applications of wireless technology are moving from any time and any place to include any thing.[11]

In short, wireless technology is spread broadly across all activities of daily life and is becoming an ever more integral and indispensable part of those activities. Reports of how the wireless revolution is changing everyday life abound in the news, and they include news of the pervasive and ubiquitous computing enabled by wireless communications, making all sorts of previously impossible things possible. These changes are driven by technological advances and by the creation of new applications that make use of those advances to provide new services and create new markets. The potential is real, but realizing it, with all of its implications for more and more wireless communications of all types, will continue to strain the spectrum management regime.

Wired Versus Wireless Communication (Propagation Versus Backhaul)

Fiber optics finally led to the demise of Grove's law, which (contrasting the remarkable rate of improvements in computing performance with the slower rate of improvements in the performance of deployed communications capabilities) forecast a doubling of the bandwidth of the telephone system every 100 years.[12] The effect of rebuilding the cable and telephone industries with an abundance of fiber-optic technology has been transformative, as has been the deployment of broadband local access infrastructure using fiber, digital subscriber line,[13] and cable modem technology. The most significant impact for wireless of the investment in this

---

[9] Andrea Goldsmith, *Wireless Communications*, Cambridge University Press, 2005.
[10] National Research Council, *Embedded, Everywhere*, The National Academies Press, Washington, D.C., 2001.
[11] International Telecommunication Union, *Internet Reports 2005: The Internet of Things*, United Nations, 2005.
[12] See, for instance, National Research Council, *Defining a Decade: Envisioning CSTB's Second 10 Years*, Proceedings of Computer Science and Telecommunications Board's 10th Anniversary Symposium, National Academy Press, Washington, D.C., 1996.
[13] Interestingly, digital subscriber line networks pose their own spectrum management challenges because wire pairs within the telephone wire plant radiate into each other.

infrastructure has been a significant reduction in the need for medium- and long-range propagation of radio-spectrum signals. In effect, wireless technology has become an important (though not exclusively) local access technique for interconnection with a huge fiber transport infrastructure for voice, data, and, increasingly, video transmission. Fiber-optic connections frequently provide these "backhaul" services, which are needed to connect distributed sites (such as cell towers) to the network. Of course, a backhaul role remains for wireless links, such as microwave and satellite communications, but the tremendous breakthrough in the cost and capacity of fiber-optic technology has shifted the focus of wireless communications more toward "last-mile" and "last-meters" issues. Another consequence is that the market in wireless services is more closely linked to the market in last-mile wireline communications services.

This shift increases the importance of wireless services that operate at shorter ranges. At the shortest ranges, near-field communication is used in such applications as touchless public transportation passes, and RFID is used for communication between, for example, vehicle transponders and tollbooths.

CHANGING MARKET DYNAMICS

Wireless technologies are making possible valuable new services and products. Most large-scale commercial applications of wireless technology have until recently operated using licensed spectrum—spectrum in which only the assigned user can operate and offer services according to the terms of its license. Broadcast television and radio, satellite communications, and cellular telephone systems are prominent examples. As personal wireless communications and related data services are improved, demand for spectrum to be used by individuals and devices continues to increase. As previously discussed, a growing number of devices (including laptops, tablets, cell phones, electronic book readers, cameras using WiFi, headsets and other devices using Bluetooth, and sensors and controls using such protocols as ZigBee) operate in open bands in which defined technical rules for both the hardware and the deployment methods are employed to enable shared use without license rights or guarantees of protection from interference. Such capabilities are being deployed by individual users (households with WiFi for sharing a broadband connection throughout their house); schools, other organizations, and firms (to provide connectivity within their premises); communications carriers (to complement their offerings using licensed spectrum or wireline connections); and local governments (for their own use or to extend communications within their communities). This complementary approach is often credited with having allowed the rapid development of new products and services. Spectrum policy, service offerings, and business models have all been evolving to take advantage of licensed operation as well as operation in open bands.

Some currently licensed spectrum uses are facing competition or replacement by technology-enabled alternatives. For instance, terrestrial broadcast television now competes with both cable and satellite transmission, and they all compete with video delivered (by streaming or download) over the Internet. Spectrum once dedicated to a particular use becomes less valuable as alternative uses become more valuable. An obvious example is the spectrum once reserved for analog television broadcasting channels and freed when broadcast television completed its transition to all-digital transmission. The question of what to do with the "white space" created by freeing spectrum previously allocated for television channels 2 to 51 has highlighted many of the arguments about the merits of licenses, the possibilities for using markets to shift spectrum to new uses, and the role of open-band approaches.[14]

Still another aspect of shifting market dynamics is related to the globalization of markets. Global markets for wireless communications devices have been driven not so much by global travelers, which are relatively few, as by the global economies of scale associated with common components, common products, and consistent standards that make it possible to develop products and services for large markets. Where differences do exist, decreasing component costs and increasing miniaturization have enabled multimode devices such as tri- and quad-mode cell phones that sidestep some of the harmonization issues.

THE EVOLVING POLICY AND REGULATORY FRAMEWORK

There appears to be a broad consensus that the current framework for spectrum policy is ripe for change.[15] This attitude reflects recognition of the shortcomings of centralized government management of spectrum use as well as the need to accommodate present and emerging technological capabilities such as those discussed in Chapter 2. A number of significant policy changes reflect efforts to adjust to new technologies and to shift some control from central management to markets and open bands. This section reviews the origins of the present policy regime and some recent efforts to make changes.

---

[14] See testimony submitted to the Federal Communication Commission, "Unlicensed Operation in the TV Broadcast Bands," ET Docket No. 04-186, and "Additional Spectrum for Unlicensed Devices below 900 MHz and in the 3 GHz Band," ET Docket No. 02-380.

[15] FCC, "Report of the Spectrum Policy Task Force," ET Docket No. 02-135, November 2002, p. 11; Government Accountability Office (GAO), *Telecommunications: Comprehensive Review of U.S. Spectrum Management with Broad Stakeholder Involvement Is Needed*, GAO-03-277, Washington, D.C., January 2003, p. 3.

History

There are several potential historiographies of the emergence of wireless communications policy in the United States. Each represents a particular perspective on the proper role for government and for markets in the management of spectrum. This section starts with a brief summary of the official administrative story—that is, the legislative and regulatory actions beginning with the Radio Act of 1912. Both the Supreme Court, when it initially upheld the role of the Federal Communications Commission (FCC) in licensing wireless systems, and the FCC in various reports (such as the Spectrum Policy Task Force report described below in this report) reflect this perspective. Three additional perspectives reflect actual or perceived motivations, priorities, and consequences from alternative points of view. Often unstated or implied in current spectrum policy debates, these stories color the assumptions and arguments made by the diverse policy stakeholders, with numerous important implications for spectrum policy analysis. They also serve to reveal the many potential pitfalls for spectrum policy making.

Official (Administrative) Story

The administrative story begins with the demise of the *Titanic* and the sense that potential rescuers could not be reached because of a lack of coordinated communications. The Radio Act of 1912 was meant to address such issues, but a 1926 court decision in *United States v. Zenith Radio Corp.* held that the 1912 act did not allow the secretary of commerce (under authority from the President) to refuse licenses.[16] That decision led to an 8-month period when the law broke down and a cacophony of signals was transmitted, so that no one could be heard, followed by the rapid passage of the Radio Act of 1927. The provisions of the 1927 act were mostly incorporated into the Communications Act of 1934, which unified the regulatory regime for nongovernmental use of spectrum for telephone, telegraph, and radio under the control of the FCC. Regulation of governmental spectrum use was assigned to the executive branch, and eventually, in the 1970s, to the National Telecommunications and Information Administration (NTIA) of the Department of Commerce. This split addressed concerns about concentrating licensing authority, as reflected in the 1926 court decision.[17] These two agencies, the FCC and the NTIA, must coordinate to accommodate the full range of spectrum users since no spectrum is specifically mandated for exclusive federal or nonfederal

---

[16] United States v. Zenith Radio Corp. et al., 12 F. 2nd 614 (N.D. Ill., 1926).

[17] GAO, *Telecommunications: Better Coordination and Enhanced Accountability Needed to Improve Spectrum Management*, GAO-02-906, Washington, D.C., September 2002, p. 2.

use.[18] The system put in place in 1934 is largely the system that we have to this day.[19]

This historiography presents spectrum management as a straightforward technical problem, to be solved to the extent possible and necessary by the most direct and straightforward regulatory mechanism.

Government Control Story

The government story starts with a focus on the Navy's efforts to control the airwaves since the early 20th century, efforts that had been almost entirely successful as the United States entered the First World War. It then follows the battle over the following decade that resulted in direct control (through the Independent Radio Advisory Committee and the NTIA) over much of wireless communications capacity, and indirect control through the private-public arrangement embodied in the FCC over the remainder. There are nuances to this story. Early versions focused on overly zealous regulation and the scarcity of capacity it caused.[20] Newer versions focus more heavily on the positive political theory (i.e., the use of game theory and other formal methods) of legislation.[21] The primary practical lessons of this perspective are that any form of regulatory solution, however well designed, can have undesired results, including corruption or failure, so that the institutional design of the regulatory system aims to minimize the role of self-conscious policy making.

Business Story

The business story focuses on the moves of the industrial players in the first quarter of the 20th century. It follows the path from Marconi to De Forest, the joining in of AT&T and later GE and Westinghouse, the formation of RCA, and the patent pools of 1920.[22] In this story, a series

---

[18] U.S. Department of Commerce, *Spectrum Policy for the 21st Century—The President's Spectrum Policy Initiative: Report 1*, June 2004, pp. 8-10.

[19] FCC, "Report of the Spectrum Policy Task Force," ET Docket No. 02-135, November 2002, p. 7. Additional source: NBC v. U.S. 319 U.S. 190, 1943.

[20] R.H. Coase, "The Federal Communications Commission," *Journal of Law and Economics* 2(October):1-40, 1959; Jora R. Minasian, "Property Rights in Radiation: An Alternative Approach to Radio Frequency Allocation," *Journal of Law and Economics* 18(1; April):221-272, 1975.

[21] Thomas W. Hazlett, "The Rationality of U.S. Regulation of the Broadcast Spectrum," *Journal of Law and Economics* 33(1):133-175, 1990; Thomas W. Hazlett, "Assigning Property Rights to Radio Spectrum Users: Why Did FCC License Auctions Take 67 Years?" *Journal of Law and Economics* 4(2):529-576, 1998.

[22] Yochai Benkler, "Overcoming Agoraphobia: Building the Commons of the Digitally Networked Environment," *Harvard Journal of Law and Technology* 11(Winter):287, 1997-1998.

of business decisions by the primary manufacturers of transmission and reception equipment in the second and third decades of the 20th century led to the emergence of the broadcast model.

Through a variety of techniques, some developed in the market, some through the patent system, and some through the regulatory system, the broadcasting industry had settled by 1926 on the advertiser-supported networks using government-granted exclusive licenses that dominated until very recently. The following years of industry consolidation saw a shift from what was primarily an equipment-market-driven phenomenon in the 1920s (e.g., the need to create demand for receivers as the economic rationale for the creation of the National Broadcasting Company) to an advertiser-supported entertainment service by the 1930s. It also saw the shift from spectrum allocation by the secretary of commerce to allocation by an independent agency, the FCC. However, the basic structure was set in place even before—and independent of—formal legislation.[23] The primary significance of perspective as a guide to contemporary policy making is in regard to the need to pay particular attention to the business structure of the markets in wireless communications equipment and wireless services and their implications for proposed institutional designs.

Public-Interest Advocates Versus Commercial Broadcasters Story

A third, and final, nonofficial story is the story of the battle between entrenched broadcasters and advocates concerned with a public interest in spectrum and publicly minded broadcast policy. In this story, much of the action that matters most occured later than in either of the two other nonofficial stories—in the period between the advent of broadcast radio and passage of the Communications Act of 1934. During that time, a variety of education, labor, religious, press, and civic groups opposed the network-based and advertising-supported system that was emerging and advocated for setting aside significant capacity for nonprofit and noncommercial broadcasting.[24] The story is important because its primary elements continue to describe a fairly broad perception of the political stakes in wireless communications policy. Broadcast communications policy is perhaps the most visible of wireless policies for most Americans.

The construct of the "public interest" evokes strong political emotions and deeply held beliefs. The political power of broadcasters, coupled with

---

[23] Erik Barnouw, *A History of Broadcasting in the United States: Volume 1: A Tower of Babel: To 1933*, Oxford University Press, New York, 1966; Hugh G.J. Aitken, "Allocating the Spectrum: The Origins of Radio Regulation," *Technology and Culture* 35(4):686-716, 1994.

[24] Robert W. McChesney, *Telecommunications, Mass Media, and Democracy: The Battle for the Control of U.S. Broadcasting, 1928-1935*, Oxford University Press, New York, 1994.

the belief that this particular area of policy is especially important for, and amenable to, political action, creates important constraints on the range of policies practically open for reform.

Allocation, Assignment, and Licensing

The allocation of frequencies for a particular use (*what* is permitted to operate in a range of frequencies) is distinct from their assignment (*who* is permitted to use that range of frequencies). Allocation was historically made through rule making; recent years have seen a shift from assignment by comparative hearing to auctions and the introduction of secondary markets to allow market-based reassignment.

The vast majority of licenses to operate wireless devices and systems in the United States are assigned in an administrative process either by the FCC, which has jurisdiction over use by private and state, local, and tribal users, or by the NTIA, which has jurisdiction over use by federal agencies.

The fundamental principal for regulation of transmitters is that it is impermissible to operate a wireless communications transmitter in the United States except by license, unless the device has very well defined technical characteristics that allow it to be operated under one of the FCC's permissive frameworks for unlicensed operation. Licenses typically include limits on the use of the equipment licensed which are typically designated in terms of the following:

- The frequency of signals transmitted by the system;
- The bandwidth of the signals;
- The power of the transmitter, given the bandwidth used;
- The antenna location and height or other design characteristics (such as direction);
- The number of other potential licensees to use equipment with equivalent characteristics; and
- The relations among licensees (e.g., license exclusivity and the presence of secondary and primary users).

Licenses typically also limit the types of services that can be offered; for example, a television band licensee cannot use that spectrum for any other use.[25]

Devices that receive and decode but cannot transmit wireless communications are not subject to the same regulatory framework (although

---

[25] The advantages of not specifying particular services are compellingly illustrated in the diversity of services that have been implemented in unlicensed bands.

some, like police radar detectors, may be regulated in other contexts). Note that because receivers contain local oscillators (to detect the signal or for their computational elements) that may interfere with other transmissions, they are subject to limits on these unintentional emissions.

Overview of Recent Policy Developments

Starting with changes made to the Communications Act in 1983, Congress has sought to encourage competition and innovation and to recognize the evolving technological reality.[26] Today, increasing use is being made of less centralized mechanisms using markets in both spectrum rights and open bands. Changes to the Communications Act authorize the FCC to collect license fees, conduct spectrum auctions, and provide for spectrum allocation flexibility.[27] Auctions have seen increasing use for making assignments, and secondary spectrum markets are emerging. The opening of new bands and the auctioning of spectrum rights, together with significant technological developments, is credited, for example with having enabled tremendous growth in the number of cell phone subscribers.

Complementing these market-based mechanisms has been growing use of open bands, in which all users are free to operate subject only to rules of the road.[28] This development had its origins in the decision to establish the so-called industrial, scientific, and medical bands at 900 MHz and at 2.4 and 5.8 GHz as open bands, an action that helped pave the way for today's widespread use of WLANs.

In recent years, two U.S. government initiatives aimed at stimulating broad reform were launched—the FCC 2002 Spectrum Policy Task Force report and associated ongoing activities, and the President's Spectrum Policy Initiative of 2004.[29]

Recent specific policy changes have included approval of ultrawideband operation, which represents a new, fundamentally different way of thinking about wireless transmission and is also the first instance

---

[26] 47 U.S.C. 157, "New Technologies and Services."

[27] FCC, "Report of the Spectrum Policy Task Force," ET Docket No. 02-135, November 2002, pp. 7-8.

[28] A variety of terms describe this approach, including "license-exempt" or "license by rule." This approach is probably most familiar as the basis for operation of WLANs, cordless telephones, and the like.

[29] FCC, "Report of the Spectrum Policy Task Force," ET Docket No. 02-135, November 2002; FCC Spectrum Policy Task Force, *Report of the Spectrum Efficiency Working Group*, November 15, 2002; U.S. Department of Commerce, *Spectrum Policy for the 21st Century—The President's Spectrum Policy Initiative: Report 1*, June 2004.

of approval for the overlay of existing services;[30] changes in licensing procedures to accommodate software-defined radios and proceedings regarding adaptive radios;[31] a decision to permit low-power devices to operate on vacant broadcast television channels;[32] issuance of a notice of inquiry for a spectrum-sharing test bed to be shared among federal and nonfederal users;[33] and adoption of rules and development of technical measures enabling the sharing of spectrum at 5 GHz between existing military radar systems and low-power unlicensed devices.[34]

Two Recent Federal Policy Initiatives

Several major federal policy initiatives were launched in recent years. These include the two described below—the FCC Spectrum Policy Task Force (and a series of proceedings that followed) and the President's Spectrum Policy Initiative—as well as the FCC National Broadband Plan that was released in March 2010.

FCC Spectrum Policy Task Force (2002)

Seeking to exploit the opportunity opened by new technological capabilities, the Spectrum Policy Task Force (SPTF) approached not only the problem of the need for changes to spectrum management and allocation but also the long-term need to allow further change to happen readily in anticipation of such technological advance. The SPTF report of 2002 introduced new models and ways of thinking about the rights of users and licensees, about the accommodation of market forces, and about the preparation for future radio technologies beyond the horizon.[35]

The FCC chair formed the SPTF in 2002 to help the FCC improve spectrum policy management in recognition of the challenges it faces to "keep pace with the ever-increasing demand for spectrum and the continuing

---

[30] FCC, Order FCC 02-48, ET Docket No. 98-153, February 14, 2002.

[31] An adaptive radio and radio technology are commonly referred to as a "cognitive radio" or a "smart radio," defined in a 2005 FCC proceeding as a radio empowered to "alter its transmitter parameters based on interaction with the environment in which it operates." See FCC, Report and Order FCC 05-57, ET Docket No. 03-108, March 10, 2005, available at http://hraunfoss.fcc.gov/edocs_public/attachmatch/FCC-05-57A1.pdf.

[32] FCC, ET Docket No. 04-186, May 13, 2004.

[33] FCC, ET Docket No. 06-89, June 8, 2006, available at http://hraunfoss.fcc.gov/edocs_public/attachmatch/FCC-06-77A1.pdf.

[34] FCC, Report and Order FCC 97-5, ET Docket No. 96-102, January 9, 1997, available at http://www.fcc.gov/Bureaus/Engineering_Technology/Orders/1997/fcc97005.pdf.

[35] FCC, "Spectrum Policy Task Force Report," ET Docket No. 02-135, November 2002, available at http://hraunfoss.fcc.gov/edocs_public/attachmatch/DOC-228542A1.pdf.

advances in wireless technology and applications."[36] The SPTF report of November 2002 sought to provide a comprehensive and systematic review of FCC spectrum policy and to catalyze reform of that policy. The report offers a number of findings and recommendations aimed at improving spectrum policy and ensuring that it is able to evolve with technology and applications.

The 2002 SPTF report summarizes the regulatory history of spectrum policy in the United States from its beginnings more than 90 years ago, covering both statutory and administrative aspects. It also notes that public interest use, such as for public safety communications and national defense, is an ongoing consideration of the regulatory process and is factored into policy decisions along with economic considerations driven by private-sector demand for services and applications. The SPTF report makes the case for spectrum policy reform, stating that the dramatic increase in demand for spectrum-based services coupled with significant and continuing technological advances makes reform not only possible but also necessary. It argues that these new and evolving dynamics are straining long-standing, outmoded spectrum policies that, unchanged, will fail to maximize the potential public benefits of spectrum-based services and applications. Specifically, it notes the potential for "smart" or "opportunistic" technology, such as software-defined radios, to allow more flexible use of spectrum. Additionally, the report notes that spectrum scarcity is of increasing concern. It refers to some evidence that allocated spectrum is being underutilized and calls for more comprehensive measurements of spectrum use to be undertaken. It sees better understanding of actual use as one means of identifying where scarcity might be mitigated through more efficient allocation and greater flexibility.

The SPTF report identifies seven key elements for a new approach to spectrum policy:

- *Maximizing flexibility of spectrum use.* A flexible-use approach to spectrum policy, in contrast to the traditional command-and-control approach, allows licensed and unlicensed users maximum autonomy to determine the highest-value use of their spectrum and allows them to make choices based on market factors.
- *Clear and exhaustive definition of spectrum rights and responsibilities.* Clarity in the rules governing use would create an environment for spectrum users to confidently negotiate alternative arrangements for maximizing value. Rules should be written to identify uses that are excluded, prohibited, or limited, allowing users to explore any options not explicitly prohibited.

---

[36] Ibid., p. 1.

- *Accounting for all dimensions of spectrum use.* Spectrum should be allocated using time in addition to traditional dimensions of frequency, space, and power. Technology advances also make possible new approaches to allocation in these traditional dimensions.
- *Promoting efficiency.* Three types of efficiency are identified: spectral, technical, and economic. There are situations where spectral and technical efficiency may take priority over economic efficiency in order to promote public interest goals. However, economic efficiency can be promoted by providing spectrum users with flexibility of use and ease of transferability. This could allow maximizing of the value of services provided.
- *"Good neighbor" incentives.* To the extent possible grouping like systems or devices (e.g., low-power systems with high sensitivity to interference) together in spectrum "neighborhoods."
- *Periodic review of rules.* Rules should be reviewed so that they can be adjusted in light of technological advances made since those rules were made. Such reviews should be scheduled at intervals that permit adjustment of business plans and investments.
- *Enforcement.* Enforcement increases in complexity with the complexity of technology and applications. Proper enforcement requires sufficient resources for monitoring use of the spectrum.

The remainder of the 2002 SPTF report focuses on approaches for avoidance of interference, alternative spectrum usage models, and promotion of access to spectrum.

First, the SPTF report addresses avoidance of interference, a problem that has been a major responsibility of the FCC from its beginning and has always been a challenge. The issues related to interference have increased in technical difficulty and prevalence due to changes created by new technology and new applications. The SPTF report argues that these changes will challenge the continued effectiveness of current approaches to managing interference avoidance. It states that a more quantitative approach to interference management should be pursued by the FCC. The SPTF report recommends that the FCC move toward assessing interference based on real-time adaptation, actual spectrum use, and interactions between transmitters and receivers rather than on transmitter operations alone, as is currently done. Control of interference could be improved by several methods other than measurement, including approaches that account for and promote receiver robustness, increased use of automated transmitter power and frequency, advanced antenna technology, tightening of out-of-band emission limits, harmonizing references to interference, developing technical bulletins explaining FCC rules regarding interference, and developing a best-practices handbook.

Second, the SPTF report examines alternative spectrum usage models.

Three models are described, including command-and-control, exclusive-use, and a commons (or open-access) model. The SPTF report concludes that spectrum policy is not generally best served by the traditional command-and-control approach but mostly requires striking a balance between the exclusive rights and commons models. The report presents the alternatives as offering a continuum over which elements of the different models may be incorporated in particular instances as necessary to best serve the public good. It identifies factors that may favor the application of one model over another depending on circumstances. Generally, the SPTF report argues that the exclusive-use model may best be applied where spectrum is relatively scarce and transaction costs associated with market mechanisms are relatively low. This contrasts with the commons model, which may best be applied where spectrum is relatively abundant and transaction costs associated with market mechanisms are relatively high. The SPTF report views the command-and-control model as best only for fulfilling compelling public-interest objectives such as conforming to treaty obligations (e.g., with respect to satellite transmissions), ensuring capacity for passive scientific observations, and supporting public safety communications. Even in these cases other models should, according to the SPTF report, be applied to the extent possible.[37]

Finally, the SPTF report recommends approaches for promoting access to spectrum, which it views as essential to continued innovation. It notes the significant market for unlicensed devices created in the relatively limited spectrum available for unlicensed use. It argues that further innovation is likely with additional available spectrum for such use. It also discusses how secondary markets involving the leasing of licensed spectrum rights might further promote access.

In each of the three areas discussed—avoidance of interference, alternative spectrum usage models, and promoting access to spectrum—the SPTF report addresses transition issues that might arise.

2008 President's Spectrum Policy Initiative

The Commerce Department has been leading an effort initiated by a presidential order to take a similar fresh look at the use and management

---

[37] It is important to note that both the market and the commons approaches claim that they would reduce spectrum scarcity. The market approach would price spectrum to clear competing uses, and the commons approach would create the conditions for markets in more intelligent devices that can successfully communicate without displacing other communications—that is, without "using" spectrum. The primary differences, then, are whether transactions costs associated with market mechanisms are higher than those associated with commons approaches (e.g., dispute resolution) and whether devices can develop the ability to clear competing uses through coordination.

of spectrum allocated to the federal government in various agencies and departments.

The resulting Federal Strategic Spectrum Plan, released in March 2008 by the NTIA, incorporates summaries of 15 agency-specific plans and integrates planning needs of the NTIA and other federal agencies.[38] The plan's aim is to support a new and evolving spectrum management system that enables more effective use of spectrum and allows dynamic access to it where feasible. According to the plan, the current system cannot readily accommodate innovations or new operational requirements. The plan states that a new model for spectrum management is required to meet the growing federal and private-sector need for spectrum. It recognizes that much of the growth will be below 5 GHz, implying that additional use must be supported in already heavily utilized spectrum space. The plan emphasizes the need for agility and an evolutionary model for spectrum management that can rapidly take advantage of technology advances, including advances in use of the various degrees of freedom. It notes that meeting the needs identified in the plan will require coordination among all stakeholders, including federal agencies, state and local public safety entities, and private-sector users as well as vendors and researchers developing and commercializing technology advances.

The plan identifies several specific future federal requirements for spectrum likely to drive spectrum policy and the methods needed to meet those requirements. First, more data and higher data rates will be needed for public safety communications and military applications, such as increased use of sensors and unmanned systems. Increased application of wireless communications for law enforcement and other federal agency needs was a common theme in agency-specific plans. Second, the demand for satellite and space-based services, including space research, global positioning systems, and remote-sensing operations for meteorological services and climate research, is expected to increase, driving the need for spectrum to support them. Use of high-frequency bands (between 3 and 30 GHz) and use of spectrum for radar and air traffic control were also identified by federal agencies as likely to grow over time. Finally, the plan noted emerging applications above 30 GHz that may drive spectrum use in this frequency range over the long term.

The 2008 plan outlines near-term and mid-term strategies for addressing federal spectrum policy needs and briefly discusses challenges and plans for developing long-term strategies. It notes that projection of future spectrum use is largely qualitative (based on anticipated require-

---

[38] U.S. Department of Commerce, *Spectrum Management for the 21st Century: The President's Spectrum Policy Initiative—Federal Strategic Spectrum Plan,* March 2008, available at http://www.ntia.doc.gov/opadhome/opad_wire.html.

ments) rather than quantitative. Recognizing that private-sector spectrum needs are also likely to grow, the plan identifies crucial improvements—automation and analytical tools, standardized generation of spectrum requirements, and spectrum forecasting methods. The near-term strategy includes 10 elements for federal use of spectrum:

- Use of commercial services where feasible;
- Smart technologies such as software-defined (cognitive) radios;
- Flexible approaches to incentives for making underutilized spectrum available to other entities;
- A range of public safety issues, including interoperability, spectrum and infrastructure sharing, and expanded microwave backhaul;
- Considerations for continuity of government;
- Improving processing time for frequency assignment requests;
- Improving methods for spectrum valuation and incentivizing economic efficiency;
- Improving technical efficiency by such methods as optimizing sharing and tradeoff analysis;
- Trend forecasting; and
- Better interagency and federal/private coordination.

The plan identifies two midterm strategies for improving spectrum management. First, it describes a unified approach to coordinating spectrum management at the federal level across the FCC, NTIA, and DOD. It also describes initial plans for creating a technology test bed to support exploration of new technologies and methods to share spectrum.

The Department of Commerce Spectrum Management Advisory Committee, convened as part of the department's spectrum policy initiative, issued a series of reports in late 2008 that examine definitions of efficient spectrum use, mechanisms for improving operational efficiency, the transition of federal services to more efficient technologies, a spectrum-sharing test bed, and federal-nonfederal spectrum sharing.[39]

---

[39] See http://www.ntia.doc.gov/advisory/spectrum/csmac_reports.html.

2

Key Technology Considerations

Radio-frequency (RF) communication saw a progression of innovation throughout the 20th century. In recent years, it has been transformed profoundly by technological advances, both in the capabilities of individual radios and in the design of networks and other systems of radios. This discussion presents some highlights of recent advances and their implications for the design of radios and radio systems and for regulation and policy. It does not aim to describe the full range of technical challenges associated with wireless communications; the interested reader is referred to the 1997 NRC report *The Evolution of Untethered Communications*,[1] which describes many of the fundamental challenges associated with wireless communications or, for a more recent view of the technology and its applications, several recent textbooks on wireless communications.[2]

---

[1] National Research Council, *The Evolution of Untethered Communications,* National Academy Press, Washington, D.C., 1997.

[2] See, e.g., Andrea Goldsmith, *Wireless Communications*, Cambridge University Press, Cambridge, England, 2005; David Tse and Pramod Viswanth, *Fundamentals of Wireless Communication*, Cambridge University Press, Cambridge, England, 2005; and Theodore S. Rappaport, *Wireless Communications: Principles and Practice*, 2nd Edition, Prentice-Hall, Upper Saddle River, N.J., 2001.

TECHNOLOGICAL ADVANCES IN RADIOS AND SYSTEMS OF RADIOS

Digital Signal Processing and Radio Implementation in CMOS

Modern communications technologies and systems, including those that are wireless, are mostly digital. However, all RF communications ultimately involve transmitting and receiving analog signals; Box 2.1 describes the relationship between digital and analog communication.

Digital signal processing (Box 2.2) is increasingly used to detect the desired signal and reject other "interfering" signals. This shift has been enabled by several trends:

- Increasing use of complementary metal oxide semiconductor (CMOS) integrated circuits (Box 2.3) in place of discrete components;
- The application of dense, low-cost digital logic (spawned primarily by the computer and data networking revolutions) for signal processing;
- New algorithms for signal processing;
- Advances in practical implementation of signal processing for antenna arrays; and
- Novel RF filter methods.

The shift relies on an important tradeoff: although the RF performance of analog components on a CMOS chip is worse than that of discrete analog components, more sophisticated computation can compensate for these limitations. Moreover, the capabilities of radios built using CMOS can be expected to continue to improve.

The use of digital logic implies greater programmability.[3] It is likely that radios with a high degree of flexibility in frequency, bandwidth, and modulation will become available, based on highly parallel architectures programmed with special languages and compilers. These software-defined radios will use software and an underlying architecture that is quite different from conventional desktop and laptop computers, but they will nonetheless have the ability to be programmed to support new applications.

High degrees of flexibility do come at a cost—both financial and in terms of power consumption and heat dissipation. As a result, the wireless transceiver portion (as opposed to the application software that communicates using that transceiver) of low-cost consumer devices is unlikely to become highly programmable, at least in the near future. On the other

---

[3] Programmability of radio functionality is distinct from the increasing degree of application programmability being introduced into mobile phones and exemplified by smart phones for which a large number of user-selected applications are available.

BOX 2.1
Analog Versus Digital Communications

In common usage, the term "analog" has come to mean simply "not digital," as in "analog wristwatch" or "analog cable TV." But for the purposes of this report it is useful to trace the meaning to its original technical usage, in early computing. From about 1945 to 1965, an era when digital computers were very slow and very costly, differential equations describing a hypothetical physical system were solved (one might say modeled) by an interconnected network of properly weighted passive components (resistors and capacitors) and small amplifiers, so that the smoothly time-varying voltages at various points in this network were precisely analogous to the time behavior of the corresponding variables (velocity, acceleration, flow, and so on) of the system being modeled. Today, we solve these same equations numerically on a digital computer, very quickly and at low cost.

In a similar way, for roughly 100 years, signals were transmitted in analog form (over wires or wirelessly) with a smoothly varying signal, representing the changing level and pitch of voice; the hue, saturation, and brightness of each point in a video image; and so forth. But just as high-speed and low-cost numerical representations and digital computations replaced analog computing, it likewise became much more reliable and less expensive to transmit digital coded numerical samples of a signal to be reconstituted at the receiver rather than to faithfully transmit a continuously varying analog representation. In digital communications, information is encoded into groups of ones and zeroes that represent time-sampled numerical values of the original (voice, music, video, and so on) signal.

Ironically, in the wireless domain, once the analog signal has been encoded into a sequence of digital values, smoothly varying forms for the ones and the zeroes must be generated so that the transmitted signal will propagate. Figure 2.1.1 shows a digital sequence of ones and zeros. The sharp on-off pulses that work so well inside a computer do not work well at all when sent through space between antennas. And so groups of ones and zeroes are represented by smooth changes in frequency, phase, or amplitude in a sinusoidal carrier, the perfect waveform of propagation. Three schemes are illustrated in Figures 2.1.2 through 2.1.4: amplitude shift keying of the carrier wave from 1 volt to 0 volts (Figure 2.1.2), frequency shift keying of the transmission frequency from f_0 to f_1 (Figure 2.1.3), and phase shift keying of the phase by 180 degrees (Figure 2.1.4). These ones and zeroes are interpreted at the receiver in groups of eight or more bits, representing the numerical value or other symbol transmitted.

continued

BOX 2.1 Continued

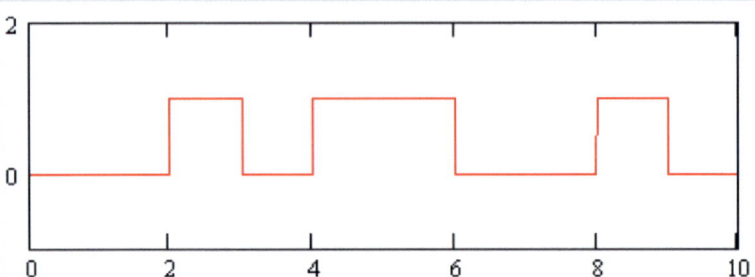

FIGURE 2.1.1 Digital sequence of ones and zeroes—0010110010. SOURCE: Charan Langton, "Tutorial 8—All About Modulation—Part 1," available at http://www.complextoreal.com. Used with permission.

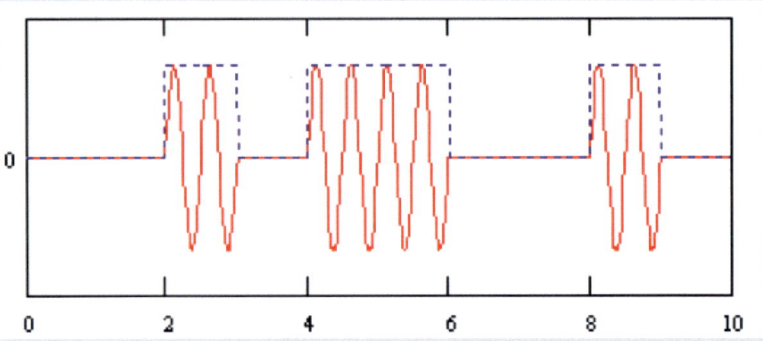

FIGURE 2.1.2 Amplitude shift keying. SOURCE: Charan Langton, "Tutorial 8—All About Modulation—Part 1," available at http://www.complextoreal.com. Used with permission.

KEY TECHNOLOGY CONSIDERATIONS

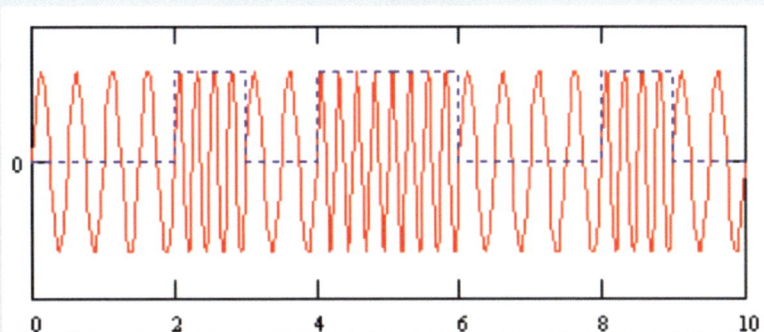

FIGURE 2.1.3 Frequency shift keying. SOURCE: Charan Langton, "Tutorial 8—All About Modulation—Part 1," available at http://www.complextoreal.com. Used with permission.

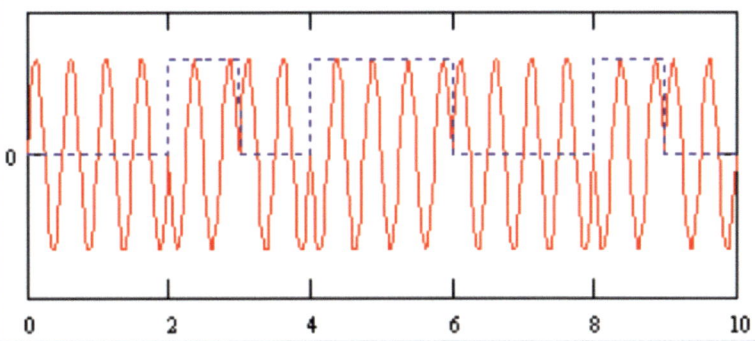

FIGURE 2.1.4 Phase shift keying. SOURCE: Charan Langton, "Tutorial 8—All About Modulation—Part 1," available at http://www.complextoreal.com. Used with permission.

BOX 2.2
Digital Signal Processing

For the continuous sinusoidal signals that can be propagated from transmitter to receiver to be encoded, modulated, demodulated, and decoded using digital technology, they must be put into a digital form by using an analog-to-digital converter (ADC), and then a digital-to-analog converter (DAC) to return to analog form. For example, an ADC might take 500 million samples per second, with a resolution of 10 bits (1 part in 1024 accuracy). Then, the continuous signal being received would be represented by a series of samples each spaced 2 nanoseconds apart. A series of dots approximately represents the continuous function shown in Figure 2.2.1.

To find the frequency domain representation of this function, we can calculate its Fourier transform. But because it is now a sequence of discrete samples rather than a continuous mathematical function, we use an algorithm known as the discrete Fourier transform (DFT). It has the form

$$X_m = \sum_{k=0}^{N-1} x_k e^{\frac{-j2\pi mk}{N}}.$$

And the inverse DFT has the form

$$x_k = \frac{1}{N}\sum_{m=0}^{N-1} X_m e^{\frac{j2\pi mk}{N}}.$$

In these two expressions, we use N time domain samples to compute N frequency components, and vice versa. A huge improvement on the DFT is the fast Fourier transform (FFT) and the inverse FFT (IFFT). By always using N equal to a power of 2 (16, 32, 64, 128…), the calculation is greatly simplified. The FFT and IFFT are the foundation of modern digital signal processing, made possible by high-speed, low-cost digital CMOS (see Box 2.3).

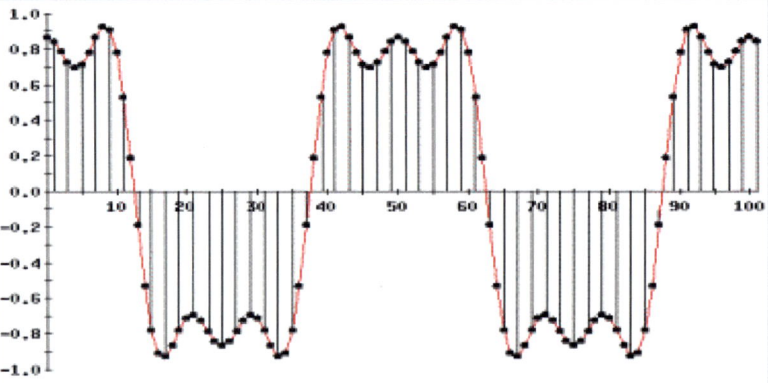

FIGURE 2.2.1 Representation of continuous function as series of digital samples. SOURCE: Charan Langton, "Tutorial 6—Fourier Analysis Made Easy—Part 3," available at http://www.complextoreal.com. Used with permission.

> **BOX 2.3**
> **Complementary Metal Oxide Semiconductor Technology**
>
> The transformation of communications from analog to digital and the related dramatic reduction in costs and increased performance are a consequence of the revolution in semiconductor design and manufacturing caused by the emergence of the personal computer (PC) industry. In particular, the remarkable and steady increase in performance and reduction in feature size by a factor of two every 18 months, generally known as Moore's law, has driven aggressive innovation far beyond the PC industry. By far, the majority investment to enable this progress has been in the design and process development of complementary metal oxide semiconductor (CMOS) technology. Introduced in the 1960s, CMOS is now used widely in microprocessors, microcontrollers, and other digital logic circuits as well as in a wide variety of analog circuits. This technology for constructing integrated circuits uses complementary and symmetrical pairs of p-type and n-type metal oxide semiconductor field-effect transistors.
>
> Investments also spawned a new industry structure: "fabless" companies, which design, market, and sell innovative products, along with silicon foundries, which manufacture the chips for these companies, spreading the capital investment in exotic equipment over large volumes.
>
> For example, today even a new, small company can design a complex part in CMOS and have a foundry charge $1,000 to process a silicon wafer yielding, say, 5,000 chips (20 cents each). Adding 10 cents for packaging and testing gives a cost of 30 cents for a part that is sold to a cell phone manufacturer for 40 to 60 cents. Well over 1 billion cell phones are sold each year.

hand, there are other applications, such as cellular base stations, where concurrent support of multiple standards and upgradability to new standards make transceiver programmability highly desirable.

Also, the decreasing cost of computation and memory opens up new possibilities for network and application design. The low cost of memory, for example, makes practical store-and-forward voice instead of always-on voice. This capability creates new opportunities for modest-latency rather than real-time communication and may be of increasing importance to applications such as public safety communications. Digital signal processing of the audio can also, for example, be used to enhance understandability in (acoustically) noisy environments.[4]

---

[4] Note that some forms of digital signal processing—compression and some algorithms used to encode and decode audio (vocoders)—can adversely affect audio quality in certain applications. For example, the vocoders in early digital mobile phones did not cope well with wind and road noise, and there have been reports that vocoders in digital public safety systems poorly transmit such important sounds as sirens and gunshots.

The pace of improvement in digital logic stands in contrast to the much slower pace of improvement in analog components. One consequence of this trend is that it becomes potentially compelling to reduce the portion of a radio using discrete analog devices and instead use digital signal processing over very wide bandwidths. However, doing so presents significant technical challenges. As a result, at least for the present, the development of radios is tied to the pace of improvements in analog components as well as the rapid advances that can be expected for digital logic, although promising areas of research exist that may eventually overcome these challenges.

Digital Modulation and Coding

Modulation is the process of encoding a digital information signal into the amplitude and/or phase of the transmitted signal. This encoding process defines the bandwidth of the transmitted signal and its robustness to channel impairments. Box 2.4 describes how waveforms can be constructed as a superposition of sinusoidal waves, and Box 2.5 describes several modern modulation schemes in use today.

The introduction of the more sophisticated digital modulation schemes in widespread use today—such as CDMA and OFDM, whereby different users using the same frequency band are differentiated using mathematical codes—have further transformed radio communications (see Box 2.6).

Many important advances have also been made in channel coding, which reduces the average probability of a bit error by introducing redundancy in the transmitted bit stream, thus allowing the transmit power to be reduced or the data rate increased for a given signal bandwidth. Although some of the advances come from the ability to utilize ever-improving digital processing capacity, others have come from innovative new coding schemes (Box 2.7).

Low Cost and Modularity

The low cost and modularity (e.g., WiFi transceivers on a chip) that have resulted from the shift to largely digital radios built using CMOS technology make it cheaper and easier to include wireless capabilities in consumer electronic devices. As a result, developing and deploying novel, low-cost, specialized radios have become much easier, and many more people are capable of doing so. A likely consequence is continued growth in the number of wireless devices and in demand for wireless communications.

BOX 2.4
Power Spectra and the Frequency Domain

Late in the 1600s, Josef Baron Fourier first proved that any periodic waveform can be represented by a (possibly infinite) sum of pure sinusoidal functions of various amplitudes. This result is surprising but true, however little the original waveform may resemble a smooth sine or cosine function. For example, a perfect square wave $x(t)$ can be represented by the infinite series

$$x(t) = \sin \omega t + (1/3) \sin 3\omega t + (1/5) \sin 5\omega t + (1/7) \sin 7\omega t + \ldots$$

Figure 2.4.1 shows that adding the waveforms of just the first four terms of this equation already begins to approximate the square wave, an approximation that improves as more terms are added.

This square wave can be composed by adding an increasing number of sine waves that are odd harmonics of the basic frequency of the square wave — that is 3, 5, 7, and so forth times the frequency — and 1/3, 1/5, 1/7, and so forth times the amplitude.

Needless to say, it is impossible in practice to combine an infinite number of sine waves, but then it is also impossible to produce a perfect square wave, rising and falling in zero time. But we certainly can generate waves with very, very fast rise and fall times, and the faster they are the larger the number of harmonics they contain. Consider just the simple case of the 3rd, 5th, and 7th harmonics.

This collection of sine waves can be represented in another way, by showing the amplitude of each frequency component visually. This amplitude spectrum (Figure 2.4.2) represents the signal amplitude in the frequency domain. A signal

continued

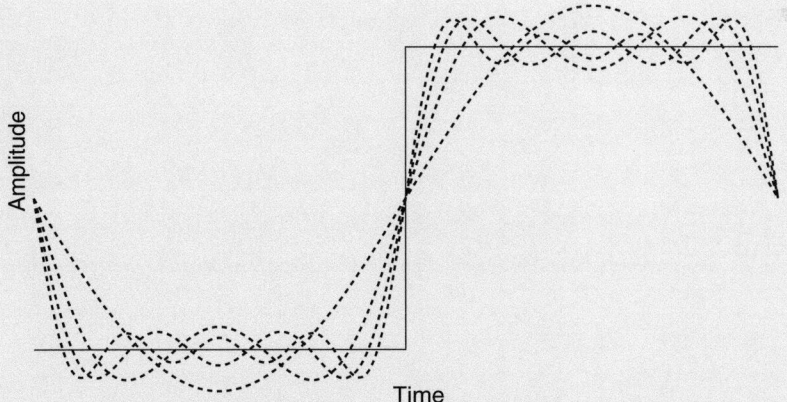

FIGURE 2.4.1 Representation of square wave (solid line) by the sum of 1, 2, 3, and 4 sinusoidal waveforms (dashed lines).

BOX 2.4 Continued

also has a frequency domain representation of the power in a signal, which is proportional to the square of the amplitude. Especially in the case of signals radiating from an antenna, we usually show the signal power spectrum as consisting of equal positive and negative frequencies or sidebands, with half of the power in each sideband. Thus, the power spectrum of the signal from Figure 2.4.1 would look like the spectrum shown in Figure 2.4.3.

These ideal-looking spectra result from combining perfectly stable, pure sine waves of precise frequencies, which are also impossible to achieve in practice. Nevertheless, the spectra do illustrate the relationship between the coefficients of the time-domain harmonics in the Fourier series, and the frequency-domain components in the amplitude and power spectra. These are more clearly related by the Fourier transform, which accepts a time domain representation of a signal, such as $x(t)$, and returns a frequency domain representation:

$$X(\varpi) = \int_{-\infty}^{\infty} x(t) e^{-j\omega t} dt.$$

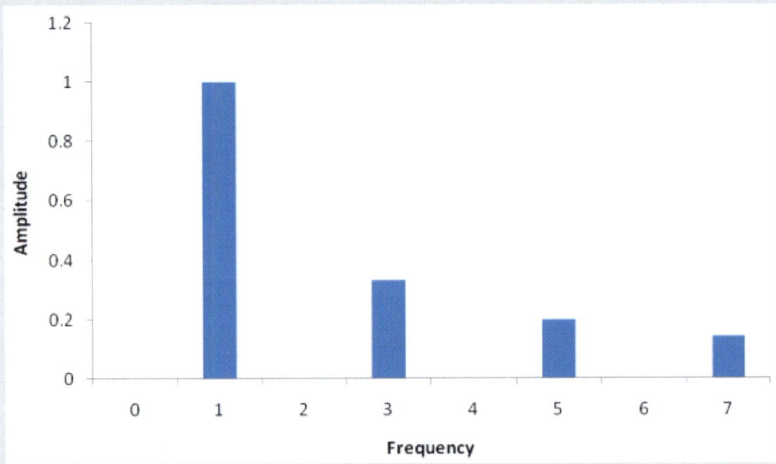

FIGURE 2.4.2 Signal amplitude represented in the frequency domain.

The inverse Fourier transform accepts a frequency domain representation $X(\varpi)$ and returns the corresponding time domain representation:

$$x(t) = \frac{1}{2\pi} \int_{-\infty}^{\infty} X(\varpi) e^{j\omega t} d\varpi.$$

These two transformations are extremely important in modern wireless, because they allow information to be encoded by including or excluding different frequencies from a transmitted signal and then detecting these at the receiver, in order to symbolize data in a way that is very resistant to interference and noise. These continuous integral equations form the basis for the discrete computations described in Box 2.2. This requires high-speed, specialized computations.

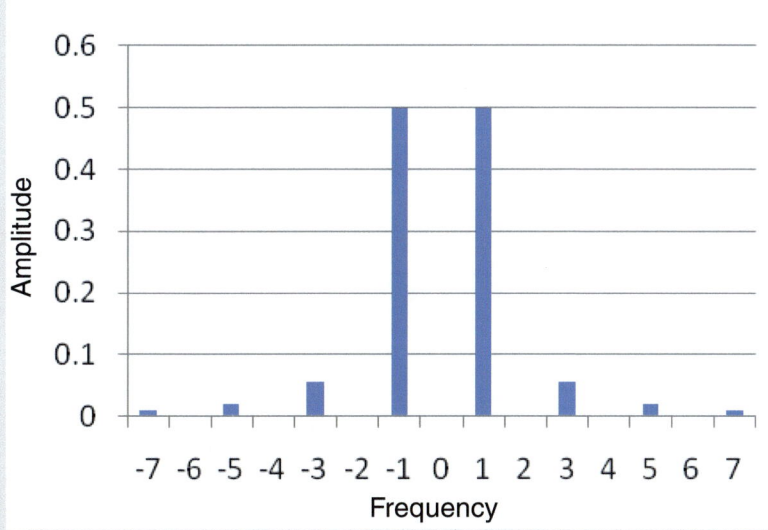

FIGURE 2.4.3 Power spectrum representation of the signal shown in Figure 2.2.1.

BOX 2.5
Modern Modulation Techniques

Gaussian minimum shift keying (GMSK) is the most widely used form of frequency shift keying (see Box 2.1), as a result of its adoption in the Global System for Mobile Communications (GSM), the standard second-generation (2G) air interface used by 80 percent of cellular phones worldwide. The sharp digital pulse used to perform the frequency shift is first softened to a Gaussian shape, reducing unwanted harmonics. The dominant worldwide cellular phone system GSM uses a simple constant-amplitude sine wave, with all modulation done by GMSK.

Quadrature phase shift keying (QPSK) is a technique that allows two bits of information to be sent concurrently. Two identical carriers 90 degrees out

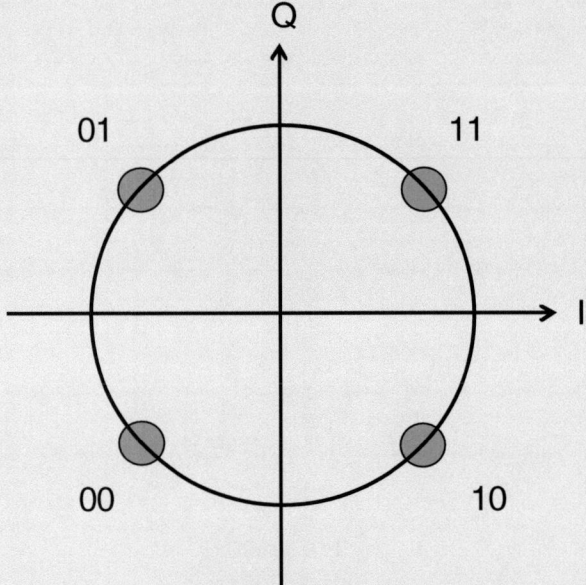

FIGURE 2.5.1 Possible pairs of bits transmitted using quadrature phase shift keying.

of phase (in-phase, *I* and quadrature, *Q*) are each modulated with a 0 degree or 180 degree phase shift to represent a one or a zero. These two modulated carriers are then added and transmitted, giving four different values, or two bits of information, when received and decoded, as shown in Figure 2.5.1.

Quadrature amplitude modulation (QAM) is a technique that, like QPSK, uses two carriers 90 degrees apart (I, Q). But instead of phase modulation, QAM uses amplitude modulation. For example, 16-QAM has four amplitude values for *I* and four values for *Q*. When the two are combined and transmitted, there are 16 possible combinations, corresponding to 4 bits, as shown in Figure 2.5.2.

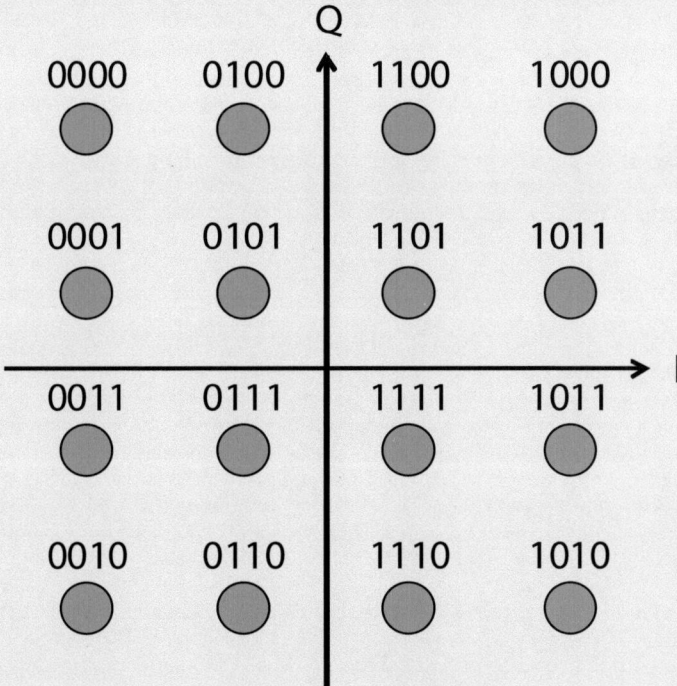

FIGURE 2.5.2 Possible groups of four bits transmitted using quadrature amplitude modulation.

BOX 2.6
Code Division Multiple Access and
Orthogonal Frequency Division Multiple Access

Because the efficient use of spectrum is important, particularly the expensively licensed cellular spectrum, it is important that as many users as possible be able to access a given frequency band without interfering with one another. Traditionally, this has been accomplished by assigning a narrowband subfrequency "channel" to each user (frequency division multiple access; FDMA), or by providing brief, high-data-rate recurring time slices or slots to each user in turn (time division multiple access). As both the number of concurrent users and the demand for higher data rates have increased, new techniques have emerged that exploit digital processing to support these requirements without increasing interference.

Code division multiple access (CDMA) has become increasingly important as the basis for higher-data-rate and more spectrally efficient third-generation (3G) mobile phone systems. CDMA is a spread spectrum multiple access technique. In CDMA a locally generated pulse train code runs at a much higher rate than the data to be transmitted. Data for transmission are added using an exclusive or logical operation with the faster-pulse train code. Each user in a CDMA system uses a different code to modulate the signal. The signals are separated by digitally correlating the received signal with the locally generated code of the desired user. If the signal matches the desired user's code, the correlation function will be high and the system can extract that signal. If the desired user's code has nothing in common with the signal, the correlation will eliminate it, treating it as noise (although it is actually rejected interference, the low correlation value makes it appear to be noise). Thus a large number of users can occupy the same band of frequencies and still be able to uniquely separate their desired signals. These coded signals use various modulation techniques such as quadrature amplitude modulation (QAM), phase-shift keying (PSK), and the like.

Orthogonal frequency division multiple access (OFDMA) uses a group of closely spaced, harmonically related subcarriers, each of which can be modulated by PSK, QAM, or other methods. They are then added and transmitted. Because the subcarriers are harmonically related, they are said to be orthogonal and can easily be separated using a fast Fourier transform and decoded. Systems often use as many as a few thousand subcarriers in a single frequency band. Rather than a particular subcarrier being assigned rigidly to each user, as in FDMA, these subcarriers can be dynamically allocated among users, providing more subcarriers to different classes of users to give higher data rates, lower error rates, or other quality-of-service choices. Also, transmitting a given payload in a given period of time over multiple channels at a lower data rate (e.g., 1 Mbps over 16 channels) is much more power efficient than transmitting over a single channel at a higher data rate (e.g., 16 Mbps over a single channel). OFDMA is the access method in the 3GPP Long Term Evolution and IEEE802.22 (WiMax) standards. The basic OFDM technique is also used in WLAN IEEE 802.11 (WiFi), Terrestrial Radio Digital Audio Broadcast, Terrestrial Digital Television (DVB-T and T-DMB standards), and Mobile Digital Television (DVB-H and ISDB-T standards).

BOX 2.7
Evolution of Coding Schemes

A major advantage of the increased use of digital CMOS technology is the ability to encode information before modulation and transmission so that errors introduced into the radio signal during transmission and reception by noise or interference can be detected and corrected during decoding at the receiver. Use of these techniques has made possible the accurate recovery of very tiny signals from heavy interference and noise. Although deep-space communication was the original source of these ideas, they have since been incorporated as a fundamental enabler of modern wireless communications ranging from wireless local area networking to mobile phones and to satellite radio and television.

For four decades, the workhorse combination making these advances possible has been the convolutional coder and the Viterbi decoder, which remain the mainstay of many systems. The convolutional coder is a simple linear state machine of shift registers and exclusive-or gates, which can be made longer and more elaborate as needed by the expected transmission environment. It expands the data prior to modulation to include additional bits that, when decoded, permit error detection and correction. The Viterbi decoder is a state machine that calculates metrics based on the current and prior received signals and makes the most likely decoding choice among the possible transmitted symbols. This scheme is the basis for CDMA and Global System for Mobile Communciations (GSM) digital cellular coding, as well as WiFi and various modems.

For the extremes of deep-space missions and certain other applications where very long codes are required, the Viterbi algorithm becomes too complex, and a more recent technique of turbo coding is combined with Reed-Solomon error-correcting codes. It is likely that turbo coding will gradually assume a central role in broadband mobile applications. In applications that are tolerant of latency, closely related low-density-parity-check codes can provide even lower error rates.

New Radio System Architectures

New networking technologies are transforming the architectures of radio systems, as seen in the introduction of more distributed, often Internet-Protocol (IP)-based networks in addition to networks that rely on centralized switching. This shift is suggested by cell phones, in which every mobile phone is a transmitter as well as a receiver, but the shift goes further to architectures that do not have the centralized control of cellular systems. What was once a population of deployed radios consisting of a small number of transmitters and many receivers (which placed a premium on low-cost receivers and did not impose tight cost constraints on transmitters) is changing to a population that contains many more transceivers. Especially in more densely populated areas, services that have

> **BOX 2.8**
> **Capacity Scaling of Mesh Networks**
>
> Much has been claimed about the scalability of mesh networks (wireless networks with no central control). Leaving aside issues of cost and latency, the possibility that they could scale linearly is enticing—that is, that adding more nodes would increase network throughput in proportion to the number of nodes added. If mesh networks were to scale linearly, they would offer infinite capacity, which would have profound consequences for spectrum management policy. However, considerable doubt has been raised about such claims by research showing limitations to their scalability.
>
> Research by Kumar and Gupta (2000) examined the question of capacity in mesh networks.[1] Their research showed that, making certain assumptions about how current technology operates, there was indeed a constraint on the ability of mesh networks to scale and that the "cause of the throughput constriction is not the formation of hot spots, but the pervasive need for all nodes to share the channel locally with other nodes."[2] It considered networks where nodes were arbitrarily located as well as networks where nodes are uniformly distributed. These results were obtained assuming perfect information about node location and traffic demand and with stationary (nonmobile) nodes. Capacity would decrease further should any of these assumptions not hold. The capacity limitations did not change when the network nodes were located in a two-dimensional plane or on the surface of a sphere. The research further showed that splitting the communication channel into several subchannels did not change any of the results. Although scaling was not linear, the results did show that networks composed of nodes with mostly close-range transactions and sparse long-range demands, such as those envisaged for smart homes, are feasible. That is, where nodes need to communicate only with nearby nodes, then

relied on transmission from high-power central sites are giving way to more localized transmissions using ever-smaller cells[5] and mesh networks (Box 2.8) that provide much greater capacity by enabling frequencies to be reused at a fine-grained level.

Dynamic Exploitation of All Degrees of Freedom

Another important shift in radios has been the ability to use new techniques to permit greater dynamic exploitation of all available degrees of freedom. Theoretic communications capacity is the product of the number of independent channels multiplied by the Shannon

---

[5] For example, cellular carriers have introduced ever-smaller cell sizes (e.g., micro, pico, femto) to optimize cost, capacity, and coverage, and broadcasters use local repeaters to extend coverage.

all nodes can transmit data to nearby nodes at a bit rate that does not decrease with the number of nodes.

Further research has explored theoretical limits of scaling wireless networks if current technological limitations could be eliminated and optimal operational strategies could be devised.[3] For instance, no assumptions were made that interference must be regarded as noise or that packet collision between nearby transmitters must necessarily be destructive. One result of this research was to show that better scaling can be achieved by different network architectures for information transport depending on attenuation. For relatively high attenuation, a multihop transport mode where load can be balanced across nodes appears to have the best scaling characteristics. For relatively low attenuation, multistage relaying with interference subtracted from the signal at each stage could theoretically attain unbounded transport capacity. Yet achieving these theoretical limits would require not only overcoming existing technical limitations but also achieving fundamental advances in information theory to understand complex modes of cooperation between nodes in a network.

---

[1] Piyush Gupta and P.R. Kumar, "The Capacity of Wireless Networks," *IEEE Transactions on Information Theory* 46(2; March):388-404, 2000.

[2] Ibid. p. 391.

[3] Liang-Liang Xie and P.R. Kumar. "A Network Information Theory for Wireless Communication: Scaling Laws and Optimal Operation," *IEEE Transactions on Information Theory* 50(5):748-767, 2004.

---

limit for a channel. In practice, the capacity (data rate) of an individual channel will be limited by the particular choice of modulation, coding scheme, and transmission power—for any particular profile of background channel noise.

Four independent degrees of freedom can be used to establish independent channels—frequency, time, space, and polarization.[6] In the past, technology and the regulatory schemes that govern it have relied principally on a static separation by frequency and space. Advances in digital signal processing and control make it possible for radios to exploit the available degrees of freedom on a dynamic basis and to coordinate their own use of the various degrees of freedom available so as to coexist with

---

[6] Polarization has seen practical application only for separating wireless signals for satellite and point-to-point microwave services.

one another and with uncoordinated spectrum occupants. Antenna arrays enable more sophisticated spatial separation through beam forming in all three dimensions. Today's radio technologies can thus, in principle, take greater advantage of all the degrees of freedom (frequency, time, space, and polarization) to distinguish signals and to do so in a dynamic, fine-grained fashion. An important consequence is that a wider set of parameters (beyond the conventional separation in frequency and space) can be used to introduce new options for allocating usage rights (i.e., defining what a user can do and what the user must tolerate) based on all of these degrees of freedom.

Flexibility and Adaptability

The agility and the flexibility of radios are improving along with advances in the ability to more accurately measure communication channels (sensing), share channels (coordination), and adapt to the operational environment in real time (adaptation). More agile radios can change their operating frequency or modulation or coding scheme, can sense and respond to their environment, and can cooperate to make more dynamic, shared, and independently coordinated use of spectrum. Digital logic advances make it possible for radios to incorporate significant and growing computing power that enables them to coordinate their own use of the various degrees of freedom available so as to coexist with each other and with uncoordinated spectrum occupants. Since much of the processing is performed digitally, the performance improvements popularly associated with Moore's law that characterize the computer industry are likely to apply to improvements in this type of processing. The result is that radios and systems of radios will be able to operate in an increasingly dynamic and autonomous manner.

Finally, increased flexibility poses both opportunities and challenges for regulators. Although it is much more complex, costly, and power consuming, flexibility makes possible building radios that can better coexist with existing radio systems. Coexistence is sometimes divided into underlay (low-power use intended to have a minimal impact on the primary user) and overlay (agile utilization by a secondary user of "holes" in time and space of use by the primary user). Such overlays and underlays might be introduced by rules requiring such changes or by rules that enable licensees to agree to such sharing in exchange for a market price.

Moreover, flexibility allows building radios with operating parameters that can be modified to comply with future policy or rule changes or future service requirements. That is, devices are able to instantiate and operate on specified policies, and the policies (and the devices' operation) can be modified.

Besides providing regulators and system operators with a valuable new tool, this malleability poses new challenges, such as how to assure a radio's security in the face of potential (possibly malicious) attempts to modify its software. Possible scenarios include rogue software silently placing calls constantly (thus congesting the control channel) or altering the parameters of a cell phone's transmitter so as to jam transmissions of cellular or other services. Information system security experience from other applications suggests that it will be possible, with significant effort, to provide reasonable security (i.e., against casual efforts to break it) but that it would be quite difficult using today's state of the art to provide highly robust security against a determined attacker.[7]

Antennas

Work has been done for many years on antennas that can operate over very wide frequency ranges. Early theoretical work in this area on mode coupling of radiation into materials by such authors as Chu,[8] Harrington,[9] and Hansen[10] still stands today, and advanced research continues on such topics as fractal and non-resonant antennas. Commercial products approximating wideband antenna technology include patch antennas, meander antennas for use at 2.4 and 5 GHz, and extreme spectrum antennas in the 2 to 6 GHz bands.

In the past decade, an interesting new approach to improved wireless communication began to develop, based on using multiple antennas at both transmitter and receiver. Advances in analog and digital processing have made it possible to individually adjust the amplitude and phase of the signal on each member of an array of antennas. When the approach is used to increase data rates, it is called multiple-input, multiple-output (MIMO), and when it is used to extend range, it is called beam forming. The most basic form of MIMO is spatial multiplexing, in which a high-data-rate signal is split into lower-rate streams and each is broadcast concurrently from a different antenna. (More generally, multiple antennas can be used to obtain the desired degree of enhancement in both data rate and range.) These schemes require significant "baseband" (i.e., digital)

---

[7] For a general discussion of cybersecurity challenges, see, for example, National Research Council, *Toward a Safer and More Secure Cyberspace*, The National Academies Press, Washington, D.C., 2007.

[8] L.J. Chu, "Physical Limitations of Omnidirectional Antennas," *Journal of Applied Physics* 19(December):1163-1175, 1948.

[9] R.F. Harrington, "Effects of Antenna Size on Gain, Bandwidth and Efficiency," *Journal of Research of the National Bureau of Standards* 64:1-12, 1960.

[10] R.C. Hansen, "Fundamental Limitations in Antennas," *Proceedings of the IEEE* 69(2):170-182, 1981.

processing before transmission and after reception, but are able to provide increased range or data rates without using additional bandwidth or power. They provide link diversity, which improves reliability, and they enable more efficient use of spectrum. This approach is used in a number of commercially deployed technologies including 802.11n (a wireless LAN standard), WiMax (a last-mile wireless local-access technology), and long-term-evolution (LTE; a technology for fourth-generation mobile telephony).

LOW-COST, PORTABLE RADIOS AT FREQUENCIES OF 60 GHZ AND ABOVE

The use of CMOS and digital processing together with other advances in RF technology opens up opportunities in the form of low-cost, portable radios that are becoming increasingly practical at frequencies of 60 GHz and above. Technological progress may extend this up to 100 GHz and beyond.

Radios operating in this domain confront a number of challenges. At these frequencies, propagation distances are very short in free space and even shorter where there is foliage. Penetration through and diffraction around building walls or other structures are also very limited. On the other hand, operation at these frequencies also has some attractive properties. Only at these frequencies are very large bandwidths available, making them the only practical option to support wireless applications that require extremely high data rates. For example, technology developed for in-room video transmission uses data rates of up to 4 gigabits per second (Gbps).

Another attractive feature of operation at these frequencies is diminished potential for interference. Short propagation distances and limited penetration of buildings are one reason. The high path losses could have another advantage with respect to interference. A likely solution to the path-loss problem is to use adaptive beam forming to provide high antenna gain—that is, directing transmitted energy along a chosen path and preferentially receiving a signal from a chosen path. If transmission sensitivity and receiver sensitivity are thus tightly focused, the potential for interference among different pairs of transmitters and receivers is markedly reduced.[11] Using these frequencies for mobile devices therefore becomes technically challenging because very narrow beams must be dynamically

---

[11] Note, however, that this situation can also exacerbate the "hidden node" problem in which a transmitter using a "listen before talk" protocol before transmitting to a receiver cannot necessarily detect another transmitter that is already using the same channel to communicate with the receiver.

KEY TECHNOLOGY CONSIDERATIONS   53

steered, a capability that is now being deployed in commercial products providing links up to 4 Gbps.[12] Note that technological advances in these areas, which open up the bands between 20 and 100 GHz to practical use, will also open up other attractive options for using antenna arrays at lower frequencies, such as the use of MIMO, or the adaptation of 802.11n technology to operate at higher-than-present frequencies.

What applications might operation in these newly accessible frequencies have? In the short term applications that require very large bandwidth over short range appear to be promising, such as devices that allow computer devices to transfer data at high speed across a desktop or devices that can transmit high-definition video from one side of a room to another.

Short propagation distances make these frequencies less viable for wide-area infrastructure or applications where in-building signal propagation is important. Looking ahead, it is possible that new architectures, such as very small cells, could make it possible to use these frequencies to provide wider-coverage services. Realizing this vision would depend on several factors not yet present—devices that operate at 20 to 100 GHz becoming cheap enough to be ubiquitous, a sufficiently widespread and cheap wired network infrastructure that would connect these devices, and the development of new business models for such services.

INTERFERENCE AS A PROPERTY OF RADIOS AND RADIO SYSTEMS, NOT RADIO SIGNALS

It is commonplace to talk about radio signals interfering with one another, a usage that mirrors the common experience of broadcast radio signals on the same channel interfering with each another. Thus, the term "interference" might suggest that multiple radio signals cancel each other out, making their reception harder or impossible. However, this view is misleading because radio signals themselves do not, generally speaking, interfere with each other in the sense that information is destroyed. In fact, interference is a property of a receiver, reflecting the receiver's inability to disambiguate the desired and undesired signals.

Radio signals are electromagnetic waves whose behavior, as described by Maxwell's equations, is linear. One consequence of this behavior is that radio signals do not, in general, cancel each other out. Each new communication signal is superposed on the entire field.[13] Actual destruction of information requires energy input at the point of destruction, and this

---

[12] For example, SiBEAM, which is currently offering chip sets for wireless high-definition television links.

[13] In principle, a precisely applied signal could literally cancel out another field, but only at a single point in space.

energy must be applied very precisely to cancel out the signal's vector field in all six dimensions, which is a low-probability event, and applies only at a single point in space.

As a result, the superposition of any number of radio signals should be thought of not in terms of destroying information but rather in terms of the ambiguity it creates for a radio trying to receive any one specific signal. The difficulty of resolving the ambiguity relates to the energy emitted by other radios (with the implication that each radio sees multiple signals) and the unpredictability of the signals (which makes the individual signals harder to separate).

Even though it is available, information is discarded in the receivers primarily for the following reasons:

- *Dynamic range*—large interfering signals inhibit a receiver's ability to detect a small signal. A small signal cannot be amplified above certain noise levels without the larger signal saturating the receiver. Moreover, desensitization circuits will reduce gain in the presence of strong interfering signals. Finally, the resolution of analog-to-digital converters is limited, which means that a weak signal cannot be digitally represented when a strong interfering signal is also present.
- *Nonlinearity of receiver components*—the desired and the interfering signals will interfere with each other inside the receiver. (See below.)
- *Inadequate separation in signal space* (within the various degrees of freedom and code).

Moreover, the extent to which signal processing can be used to separate signals with the required sensitivity, accuracy, and latency is limited by the computational power available in a radio. Removing signal ambiguity thus entails investment in one or more of the following: better radio components, additional radio complexity, additional integrated circuit area, additional antennas, additional computation, and/or additional power consumption.

Another area for potential improvement is in systems of radios. With more and more radios capable of transmitting and receiving, behavioral schemes can be used to mediate among radios. Also, because it is fundamentally easier to separate out known (and thus predictable) signals as opposed to random signals, mechanisms that allow waveform and modulation information to be registered or otherwise shared may prove useful. However, there will always be practical limits to what can be shared or coordinated.

The costs of disambiguating signals are, thus, ultimately reflected in a number of ways, including in the complexity of a radio's (or system's) design, the cost of its hardware, its size, the power it consumes, and (for mobile devices) the lifetime of the battery. Disambiguation thus involves tradeoffs, given that a radio is built to meet many requirements, only one of which is dealing with signal ambiguity.

ENDURING TECHNICAL CHALLENGES

Even as the capabilities and performance of radios continue to improve, a number of hard technical problems can be expected to persist.

Power Consumption

The power required to operate increasingly complex and sophisticated radios will continue to represent an important boundary condition, especially for mobile devices, where it dictates the cost, capacity, dimensions, and weight of their batteries as well as the interval between charges. Even for nonmobile devices, excess power results in heat that requires space or costly cooling components to dissipate. The design of practical radios will continue to reflect difficult tradeoffs between power consumption and other desired attributes and capabilities.

Nonlinearity

Real-world radio elements are not perfectly linear—that is, the output of an element is not exactly proportional to the input. Nonlinearity results in signal distortion and, when more than one signal is present in a nonlinear element, the creation of new, unwanted products of the original signals—an effect known as intermodulation distortion. The result is a degraded ability to separate a desired signal from other signals, which constrains the extent to which a receiver can mitigate interference.

Radio designers use several strategies to mitigate these effects. One is to use filters that separate out signals at other frequencies from the range of signals that are to be detected. In particular, filters allow relatively strong signals to be separated out so that a relatively weak signal can be detected. Another is to use components that are close to linear over a wider range of signal strengths.

Nonlinearity has always been a significant challenge to radio designers. It is a particular challenge to realizing the vision of radios that dynamically adapt to the presence of other radios by changing their frequency and other operating parameters. One might imagine building a radio that uses digital signal processing over very wide frequency ranges to separate out desired signals from potentially interfering signals. Doing so would allow one to leverage improvements in digital logic and better digital signal processing techniques to mitigate interference. However, the extent to which this approach can be used is constrained by the intermodulation distortion associated with real-world radio components, which limit the bandwidth that can be handled practically using digital signal processing alone.

A variety of avenues are being pursued by researchers to overcome these constraints. One of them has long been of interest but has not been realized in commercial products: the use of narrow filters that are tunable under digital control over a wide range, perhaps using microelectromechanical systems (MEMS) technology.

Nomadic Operation and Mobility

Supporting nomadic operation and mobility requires more dynamic adaptation of radio operating parameters than is needed for fixed radios, which only need to cope with changes in environmental conditions. Moreover, nomadic operation and mobility make it more difficult to neatly segment space or frequency, and they complicate dynamic market approaches because they make it more difficult to buy and sell rights at the rate at which radios can move between segments.

Heterogeneity of Capabilities

As more sophisticated radios are deployed, the heterogeneity of capabilities—especially the existence of radios with much poorer performance than others—will present growing challenges. At any point in time, there will be a legacy in terms of deployed equipment, existing frequency allocations, and existing businesses and government operations that are being made obsolete, in some sense, by new capabilities. The problem is not new, but a rapid pace of technological advancement and concomitant explosion of applications, especially applications with different purposes and capabilities, magnifies the challenges.

Not all heterogeneity will arise from legacy systems. Some applications will have cost and/or power requirements that preclude the use of highly sophisticated radios that coordinate their behavior. For example, the constraints on cost and power consumption for embedded networked sensors preclude the use of highly sophisticated radios that are able to do very sophisticated signal processing or complex computation to coordinate their behavior. Another manifestation of heterogeneity is the contrast between active use, which involves both transmitter(s) and receiver(s), and passive spectrum use (e.g., remote sensing and radio astronomy), which involves receivers only.[14] Figuring out how to simultaneously

---

[14] For a detailed discussion of passive scientific uses, see National Research Council, *Frequency Allocations and Spectrum Protection for Scientific Uses*, The National Academies Press, Washington, D.C., 2007.

accommodate more sophisticated and adaptable radios with those that are necessarily less sophisticated will be an ongoing challenge.

TIMESCALES FOR TECHNOLOGY DEPLOYMENT

A particular challenge in contemplating changes to policy or regulatory practice is determining just how quickly promising new technologies will actually be deployable as practical devices and systems and thus how quickly, and in what directions, policy should be adjusted.

Rate for Deployment of New Technologies as Practical Devices and Systems

As is natural with all rapidly advancing technology areas, concepts and prototypes are often well ahead of what has been proven feasible or commercially viable. The potential of adaptive radios, for example, has been explored (particularly for military use), but the technology has not yet been used in mainstream commercial devices or services. As described above, there is reason to expect the capabilities of radios to improve and their hardware costs to steadily decline, but many important details of operation and protocols must be worked out in parallel with technical development and regulatory change. Moreover, although great technical progress has been made in recent years, resulting in the deployment of new wireless services, wireless communications will remain a fertile environment for future basic research as well as product and service development.

Timescales for Technology Turnover

Different wireless services are characterized by the different timescales on which technology can be upgraded. The factors influencing the turnover time include the time to build out the infrastructure and the time to convince existing users (who may be entrenched and politically powerful) to make a shift. For instance, public safety users tend to have a long evolution cycle, as government procurement cycles are long and products are made to last a long time. Cellular turnover is rapid by comparison, and technology can be changed out relatively readily (a 2-year handset half-life and a 5- to 7-year time frame for a shift to new technology are typical). The digital television transition that finally occurred in the United States in 2009 is emblematic of the challenge of making a transition where technology turnover is very slow, in part because of expectations raised by static technology and services that were developed over many decades.

Importantly, the rate at which turnover is possible depends on the incentives for upgrading as well as the size of the installed base. For instance, firms operating cellular networks have demonstrated an ability to upgrade their technology fairly quickly despite having an enormous user base, whereas aviation has a relatively small set of users but a very long turnover rate, having yet to transition from essentially 1940s radio voice technology. The primary driver of successful upgrades is for users to see tangible benefits and for service providers to have an incentive to push for the switch. Cellular subscribers gain tangible benefits from newer capabilities commensurate with the added costs. (Also, U.S. mobile operators generally subsidize handset cost, because it makes it easier to upgrade their network technologies and increase system capacity, somewhat offsetting the visible costs to the end user),[15] whereas private pilots would incur a large capital cost and have to learn a new system even though the existing technology already meets their requirements.

TALENT AND TECHNOLOGY BASE FOR DEVELOPING FUTURE RADIO TECHNOLOGY

The changing nature of radios is creating new demands for training and education. Research and development (R&D) for radios depend on skills that span both the analog and the digital realms and encompass multiple traditional disciplines in electrical and computer engineering. Similarly, making progress in wireless networks often requires expertise from both electrical engineering and computer science. It is thus not straightforward for a student to obtain the appropriate education and training through a traditional degree program. The nature of modern radios presents another barrier to advanced education and university-based research, because the CMOS chips that lie at their heart require very large-scale fabrication facilities, presenting a significant logistical barrier to university-based groups that seek to test and evaluate new techniques.

This report assumes a continued stream of innovation in radio technology. Such sustained innovation depends on the availability of scientific and engineering talent and on sustained R&D efforts. Considerable attention has been focused in recent years on broad concerns about the declining base of scientific and engineering talent and levels of research support in the United States and its implications for competitiveness, including in the area of telecommunications. For a broad look at trends and their implications for science, engineering, and innovation, see *Rising*

---

[15] Incentives may differ across markets and regulatory regimes. For example, cellular upgrades have been market-driven in the United States and government-driven in the European Union (EU). The effect has been mixed. On the one hand, the EU push for third generation arguably got ahead of actual market demand, whereas the U.S. market moved slowly from analog to second-generation digital services, arguably giving the EU higher-quality wireless voice services sooner.

KEY TECHNOLOGY CONSIDERATIONS 59

Above the Gathering Storm: Energizing and Employing America for a Brighter Economic Future;[16] for a study focused on telecommunications research, see *Renewing U.S. Telecommunications Research*.[17]

The issues and opportunities described in this report involve considerations of many areas of science and engineering—including RF engineering, CMOS, networking, communications system theory, computer architecture, applications, communications policy, and economics. Addressing the challenges and realizing the opportunities will require a cadre of broad systems-oriented thinkers. Building this talent will be a major national advantage.

Radio engineering is an important area for consideration in this context, given that wireless is a fast-moving, high-technology industry that is economically important in its own right and that has much broader economic impacts. Moreover, wireless engineering encompasses an extensive skill set—including RF engineering, an ability to do RF work in CMOS technology, and an ability to work on designs that integrate RF and digital logic components—that is difficult to learn in a conventional degree program. Similarly, wireless networks involve expertise that spans both electrical engineering and computer science.

Finally, for R&D to be effective, it is important to be able to implement and experiment with new ideas in actual radios and systems of radios. Work on new radio designs requires access to facilities for IC design and fabrication. Work on new radio system architectures also benefits from access to test beds that allow ideas to be tested at scale. Given the high cost of such facilities, university R&D can be enhanced by collaboration with industry.

MEASUREMENTS OF SPECTRUM USE

The standard reference in the United States for the use of spectrum is the U.S. Frequency Allocation Chart that is published by the NTIA. The chart separates the spectrum from 30 MHz to 300 GHz into federal or nonfederal use and indicates the current frequency allocations for a multitude of services (cellular, radiolocation, marine, land mobile radio, military systems, and so on).

Although this chart is an invaluable reference in providing a comprehensive view of what frequencies are potentially in use for various

---

[16] National Academy of Sciences, National Academy of Engineering, and Institute of Medicine, *Rising Above the Gathering Storm: Energizing and Employing America for a Brighter Economic Future,* The National Academies Press, Washington, D.C., 2007.

[17] National Research Council, *Renewing U.S. Telecommunications Research,* The National Academies Press, Washington, D.C., 2007.

services and in giving some indication of the complexity of frequency use, it does not shed light on a particularly critical issue—the actual density of use of the spectrum. That is, are there blank spaces in frequency, time, and space that could potentially be used for other purposes?

It is increasingly asserted that much spectrum goes unused or is used inefficiently. Yet relatively little is known about actual spectrum utilization. Licensees and users are not required to track their use of spectrum. There are no data available from any sort of ongoing, comprehensive measurement program. And when spectrum measurements have been made, they were often aimed at addressing a specific problem. Proxy measurements, such as the number of licenses issued in a frequency range, have been used to characterize trends and extrapolate likely use, but they do not measure actual use and do not, of course, yield any insight into unlicensed use.[18]

Why Spectrum Measurement Is Hard

Perhaps the greatest challenge is that any program of measurement will be limited in its comprehensiveness if all the degrees of freedom are actually to be measured. Measurements can be made only at specific locations and times; measurements at one place may not reveal much about even nearby points. Results obtained by one set of measurements are not easily applied to a different situation. The full scope of measurement is suggested by the electrospace model, in which one specifies the frequency, time, angle of arrival (azimuth, elevation angle), and spatial location (latitude, longitude, elevation) to be measured.[19] Other measurement considerations include polarization, modulation scheme, location type (e.g., urban, suburban, or rural),[20] and which signals are being measured (known signals, unknown signals, or noise).

Many radio systems are designed to operate with very low average power levels, and naive spectrum measurement techniques may miss use by such low-power devices.[21] Moreover, a directional signal will be missed if the receiver is not pointing in the right direction. Often designed to operate with very low average power levels, point-to-point microwave

---

[18] Robert Matheson, *Spectrum Usage for the Fixed Services*, NTIA Report 00-378, March 2000, p. xi.

[19] Robert Matheson, "The Electrospace Model as a Frequency Management Tool," Addendum to the Proceedings of the 2003 ISART Conference, 2003.

[20] Allen Petrin, "Maximizing the Utility of Radio Spectrum: Broadband Spectrum Measurements and Occupancy Model for Use by Cognitive Radio," Ph.D. Thesis, Georgia Institute of Technology, August 2005, p. 6.

[21] Robert Matheson, letter to David Liddle in follow-up to presentation to the committee, August 27, 2004.

links and radar systems are examples of use that may be missed by spectrum measurements efforts. Radar emits narrow high-power pulses infrequently, making them easy to miss. Some uses, such as public safety communications, are inherently sporadic and random in time and location. Because they are normally confined to military installations, defense uses may take place in well-defined locations but will vary considerably over time.

Also, measurements by definition measure only active use of the spectrum; passive use of the spectrum and remote sensing cannot be detected and, worse, could be interpreted as non-use of parts of the spectrum that would be seen as empty. Similarly, without careful interpretation, guard bands established to mitigate interference for existing services could be interpreted as unused portions of the spectrum even though these bands are in a real sense being used to enable those services.

These considerations suggest that spectrum measurement is a challenging endeavor that requires measurements at many points in space and time and the collection of a very large amount of data. They also suggest that spectrum measurement has an inherent element of subjectivity, because results may depend significantly on the particular assumptions made and methods employed.

Looking forward, measurement might be improved over the long term by requiring systems to provide usage statistics, as might the development and adoption of a formal framework for measuring, characterizing, and modeling spectrum utilization. Such a framework might provide researchers a way to cogently discuss spectrum utilization and provide policy makers with evidence-based information about technical factors affecting efficient utilization. [22]

Results from Some Measurement Activities

The NTIA has a long history of spectrum measurement work going back to at least 1973.[23] Those early efforts included federal land mobile radio measuring use in the 162-174 and 406-420 MHz range, and Federal Aviation Administration radar bands in the 2.7-2.9 GHz range. These projects were generally considered successful because the measurements focused on a definite problem and were able to address specific questions, such as whether claimed interference was real and whether minor changes to receivers could mitigate the problem of overcrowded use. The

---

[22] F. Weidling, D. Datla, V. Petty, P. Krishnan, and G.J. Minden, "A Framework for R.F. Spectrum Measurements and Analysis," *Proceedings of IEEE Symposium on New Frontiers in Dynamic Spectrum Access Networks,* 2005, pp. 573-576.

[23] Ibid.

NTIA conducted a number of broadband spectrum surveys in different cities in the 1990s.[24]

An NTIA report from 1993 (and updated in 2000) used proxy information as a "measurement" of spectrum usage for fixed services (e.g., common carriers).[25] That report examined historical license data and observations about market and technology factors likely to affect spectrum use, in order to gain insight on the degree to which the existing fixed-service spectrum bands would continue to be needed for their allocated services. One conclusion to be drawn from that report is that point-to-point microwave bands are probably underused and that the growth expected when these bands were allocated decades ago did not occur.[26] Anticipated use of point-to-point microwave has moved largely to optical fiber instead, although it is still used in many rural areas where the traffic does not justify the cost of laying fiber.

A number of research projects have attempted to directly measure spectrum utilization.[27] Shared Spectrum Company, a developer of spectrum-sensing cognitive radio technology, has made several measurement studies since 2000, including occupancy measurements in urban settings such as New York City and Chicago, suburban settings such as northern Virginia, and rural environments in Maine and West Virginia.[28] Spectrum measurements for the New York City study were done during a period of expected high occupancy, the Republican National Convention.[29] The studies aimed to determine how much spectrum might be allocated for more sophisticated wireless applications and secondary users relative to primary (licensed) users.

Some important conclusions can be drawn from these measurements. The measurements indicate that some frequency bands are very heavily

---

[24] Frank H. Sanders and Vince S. Lawrence, *Broadband Spectrum Survey at Denver, Colorado,* NTIA Report 95-321, September 1995; Frank H. Sanders, Bradley J. Ramsey, and Vincent S. Lawrence, *Broadband Spectrum Survey at San Diego, California,* NTIA Report TR-97-334, December 1996; Frank H. Sanders, Bradley J. Ramsey, and Vincent S. Lawrence, *Broadband Spectrum Survey at San Francisco, California,* May-June 1995, NTIA Report 99-367, July 1999.

[25] Robert Matheson, *Spectrum Usage for the Fixed Services,* NTIA Report 00-378, March 2000, p. 1.

[26] Robert Matheson, letter to David Liddle in follow-up to presentation to the committee, August 27, 2004.

[27] P.G. Steffes and A.J. Petrin, "Study of Spectrum Usage and Potential Interference to Passive Remote Sensing Activities in the 4.5 cm and 21 cm Bands," *Proceedings of the IEEE Geoscience and Remote Sensing Symposium* 3(20-24):1679-1682, 2004; S.W. Ellingson, "Spectral Occupancy at VHF: Implications for Frequency-Agile Cognitive Radios," *Proceedings of the IEEE Vehicular Technology Conference* 2(25-28):1379-1382, 2005.

[28] Mark McHenry, "NSF Spectrum Occupancy Measurements Project Summary," Shared Spectrum Company, August 15, 2005.

[29] Mark McHenry and Dan McCloskey, "New York City Spectrum Occupancy Measurements September 2004," Shared Spectrum Company, December 15, 2004.

used and that some other currently assigned frequency bands are only lightly used, at least over some degrees of freedom. Above all, the picture that emerges clearly from the measurements made to date is that frequency allocation and assignment charts are misleading in their suggestion that little spectrum is theoretically available for new applications and services—provided that the right sharing or interference mitigation measures could be put in place. One might legitimately quibble over the details or the precise level of use; the real point is that there is a good deal of empty space, provided that ways of safely detecting and using it can be found.

Another broad conclusion is that the density of use becomes lower at higher frequency. The advent of low-cost radios that can operate at frequencies in the tens of gigahertz points to a promising arena for introducing new services.

Finally, measurements of spectrum use do not capture the value of use. In addition, if a licensee internalizes the opportunity cost of underutilized spectrum and has a way to mitigate that cost, there is no need for centralized measurement and management; that empty space exists, but the best way to use it is not necessarily for the government to allow additional users.

CHALLENGES FACING REGULATORS

Technology advances bring new issues before regulators that require careful analysis. Some require a subtle understanding of the ways in which new technology may necessitate new regulatory approaches and a challenging of past assumptions about limitations and constraints. Several examples are discussed below.

Use of White Space to Increase Spectrum Utilization

The basic goal of "white space" utilization is to let operators with lower priority use the space when higher-priority users leave the spectrum unoccupied. From a technical perspective this approach requires adding sensing capability to devices to determine if a higher-priority user is using the spectral band (or bands). Such a dynamic use of spectrum has not been supported in past regulatory models.

In the dynamic situation envisaged in the white-space model, several new questions and considerations have to be addressed. For instance, "occupancy" must be defined thoughtfully. Higher-priority users opposed to the use of white space might say that any use of their spectrum could cause harm to their transmissions, so that only "no interference" is acceptable. Yet achieving no interference has never been possible because all radios transmit energy outside their allowed bands and generate interference with other adjacent users. Therefore, the only question, ultimately, is the degree to which interference is allowed. In the absence of a clear technical analysis of when a given level of interference is actually causing significant degradation of signal, it is difficult to determine an acceptable level. How best to do so is of importance in formulating rules to open up spectrum as well as for private parties to negotiate what level of interference they would accept in return for a market price.

A clear technical analysis requires that several factors be considered. Estimating the total interference load depends on a realistic statistical model for the number of likely secondary users, the transmitted power spectrum for each user, the susceptibility of the primary occupant's receivers to these secondary signals, and the ability of the primary user to adapt its transmissions to reduce the impacts of the secondary users.

Given that the analysis is statistical in nature, it may be useful to approach the question in terms of a probability of degradation that should not be exceeded. If the likelihood of degradation by secondary users falls below this probability, then those secondary users would be considered as not occupying the band of the primary user. An analysis done from this perspective would help avoid situations in which highly improbable scenarios (as opposed to situations that can reasonably be expected to cause a problem) lead to the rejection of sharing arrangements.

Second, considering frequency as the only degree of freedom available to separate users makes for simpler technical analysis but is highly limiting. Radios built to perform dynamic beam forming, for instance, allow highly sophisticated spatial separation. Also, if sensing is fast enough, then it is possible to exploit white spaces in time. Thus frequency, time, and space could all be considered as tools to reduce the effects of interference to below the level of degradation defined as noninterference.

Third, spectral emissions regulations have historically considered each transmitter working independently. Yet, considering sensing performed by the network might mean much greater opportunity for more efficient spectrum use. Just how much might be gained from such an approach is not well understood, because it depends on an understanding of the statistical correlation between sensing at different locations. Considering such an approach requires the same mind-set change as described previously, which allows for statistically based improvements.

Finally, there is the issue of sensitivity of detection. Greater sensitivity increases the probability of detection but also leads to a greatly increased probability of false alarms. In other words, at some point increasing sensitivity causes any random noise to appear as occupancy. To make a proper analysis requires a level of understanding about sensing that goes beyond just sensing the energy in a spectral band. Most signals have distinctive signatures that can be used to differentiate them from noise or other spurious emissions.

One opportunity to make use of white space is in the broadcast television bands. To that end, in late 2008, the FCC issued a set of rules[30] under which devices use geolocation and access to an online database of television broadcasters together with spectrum-sensing technology to avoid interfering with broadcasters and other users of the television bands. (Alternatively, the ruling provides for devices that rely solely on sensing, provided that more rigorous standards are met.) Debate and litigation ensued following the 2008 order on such issues as how to establish and operate a database of broadcaster locations. In a second order issued in 2010 to finalize the rules, the requirement was dropped that devices incorporating geolocation and database access also must employ sensing.[31]

Adaptive Antenna Arrays and Power Limits

Antenna arrays at transmitters and receivers are being used increasingly to provide greater range, robustness, and capacity. Yet the basic regulatory strategy of defining an equivalent isotropically radiated power level for transmitters ignores many of the special characteristics of antenna arrays. As one example, this regulatory approach does not encourage the use of beam forming, which has considerable advantages in reducing interference over omnidirectional antennas.

Decreasing Cost of Microwave Radio Links

The present report describes above how standard CMOS technology can now be used to transmit in the microwave bands (60-GHz links have been demonstrated). As desired data rates rise into the gigabit-per-second range, adaptive antenna arrays will be used to obtain the necessary received power for both mobile and fixed devices. As with the previous examples, this technology is very different from what has been in use until now.

---

[30] FCC, "Second Report and Order and Memorandum Opinion and Order in the Matter of Unlicensed Operation in the TV Broadcast Bands," ET Docket No. 04-186, and "Additional Spectrum for Unlicensed Spectrum for Unlicensed Devices Below 900 MHz and in the 3 GHz Band," ET Docket 02-380, FCC 08-260, Washington, D.C., November 14, 2008.

[31] FCC, "Second Memorandum Opinion and Order in the Matter of Unlicensed Operation in the TV Broadcast Bands," ET Docket No. 04-186, and "Additional Spectrum for Unlicensed Spectrum for Unlicensed Devices Below 900 MHz and in the 3 GHz Band," ET Docket 02-380, FCC 10-174, Washington, D.C., September 23, 2010.

ENGINEERING ALONE IS OFTEN NO SOLUTION

The previous section describes several specific issues where engineering insights would help to inform future policy and regulation. At the same time, it is important not to oversell the extent to which better engineering or understanding of the technology alone can yield solutions. In the end, an engineering analysis depends on a knowledge of possible scenarios and what the acceptable outcomes are. These inform a complex set of business, marketing, and political judgments about value and risk. For example,

- Engineering alone does not determine whether a service supporting aviation merits greater protection from interference than a service delivering entertainment.
- The density and the distribution of a constellation of mobile devices (which affect their ability to interfere) cannot be determined fully a priori. They will reflect market and consumer behavior, and moreover they will change over time.

3

Policy Options

PRESSURES ON TODAY'S WIRELESS POLICY FRAMEWORK

The current wireless policy framework is based on the technology of more than 8 decades ago and on the desire, at that time, for governmental control over communications. It has evolved to encompass a patchwork of legacy rules and more modern approaches that have been added over time. Nonetheless, there is wide acceptance that the rules are ripe for change, to better reflect the technological options available today and in the future. The current framework is under pressure today on several fronts:

- *The current framework continues to rely heavily (with a few exceptions) on service-specific allocations and assignments that are made primarily by frequency band and geographic location and does not encompass all of the spectrum management approaches possible.* Allocation and assignment of services by frequency band were historically seen as the only technologically feasible way of allowing multiple wireless systems and services to coexist. Today, technology advances make it possible to use additional degrees of freedom to separate transmissions, introducing new options for allocating usage rights. In addition, new frontiers are being opened by the emergence of inexpensive, small devices that operate at 20 to 100 GHz.
- *Despite revisions aimed at ensuring greater flexibility, the current framework continues to rely significantly on centrally managed allocation and assignment, with government regulators deciding how and by whom wireless communications are to be used.* Spectrum policy has become more flexible over the past several decades in such areas as permitted modulation waveforms and types

of use and the adoption of less centralized models such as unlicensed bands and white space. Nonetheless, the past decade has seen widespread agreement that central management by regulators is inefficient and insufficiently flexible—an agreement that reflects the complexity of the problem and the dispersion in the economy of the information that is required to make decisions.[1] It also reflects concerns about whether government institutions are sufficiently nimble to make efficient and timely decisions.

- *The current framework will not be able to satisfy the increasing and broadening demand for wireless communications.* One source of this demand is greater use of richer media (such as video) that requires higher data rates. Another is the continued growth in Internet applications and services and the growing demand for untethered and mobile access to them. Demand for mobile access to the public telephone network has continued—the leading example of a more general shift toward mobile interpersonal communication. Together, these have resulted in rapid growth in the number of users of wireless devices and services. Increasingly, communications are between devices as well as people, notably reflected in growing interest in sensor networks, and together, these trends may overwhelm the ability of the existing framework to enable introduction of new communications services to meet demand.

- *The current framework does not fully reflect changes in how radios are built and deployed now or in how they could be built and deployed in the future in response to different regulations.* Technological innovation has expanded the range of potential wireless applications and services and the technical means for providing them. At the same time, it has dramatically lowered the cost of including wireless capabilities in devices. The old regime and technology placed a premium on simple, low-cost receivers and did not impose tight cost constraints on transmitters. New technology enables the deployment of many more, and more capable yet inexpensive, transceivers. Today, the population of deployed radios has shifted from one dominated by a small number of transmitters and many receivers to a population that also contains many more transceivers (e.g., every cell phone is a transmitter as well as a receiver).

KEY CONSIDERATIONS FOR A FUTURE POLICY FRAMEWORK

Enabling More Nimble Evolution of Spectrum Policy

The current spectrum plan reflects decades of historical practice and in its myriad allocations and assignments reflects many stages of tech-

---

[1] For example, this point arose repeatedly in the remarks of those who briefed this committee.

nology and policy development. It thus encompasses not only the fixed allocations made years ago for services that use now-outdated technology (e.g., AM radio) but also new regulatory and technology approaches (e.g., ultrawideband).

Generally speaking, allocations for services reflect the frequency range that was practical at the time a particular service was introduced, and many services introduced decades ago persist today. It was often possible to fulfill demand for new services by exploiting the higher frequencies that would become available as advances were made in device and radio technology. Today, propagation and penetration considerations constrain, for many applications, the utility of the higher frequencies that are less crowded, and so it is no longer possible to free up spectrum for those applications by simply moving to higher frequencies. In addition, following many decades of fitting in new services wherever there was free space, there is little or no unclaimed spectrum at lower frequencies. Much of the current spectrum frame work also reflects a time when operating rights were fairly well defined and when there were relatively few systems, system operators, and transmitters.

The complexity and density of existing allocations, assignments, and uses, coupled with competing demands for new uses, especially at lower frequencies, mean that any change will be difficult. It will involve careful consideration of the specifics of allocations, assignments, and uses in specific frequencies as well as the particular technical characteristics of particular frequencies and proposed applications. Regulators must approach this evaluation carefully lest they end up simply reinventing old command-and-control approaches. Change may also involve addressing the costs and benefits of proposed changes that are (often unevenly) distributed over multiple parties, resolving conflicting claims about costs and benefits, and addressing coordination issues, which are especially challenging if achieving a particular change requires actions by a large number of parties. Moreover, some parties gain by changing while others gain by waiting. As a result, decision making ends up, broadly, being a political question.

Today, more flexible and adaptable radios and a world in which these systems are in the hands of millions of people suggest the need for correspondingly nimble and flexible processes for developing and evolving future wireless communications policy. In essence, what will be needed is an approach that is not necessarily completely right the first time, but right over time. That is, the approach should allow for experimentation and feedback, and the regulatory system should be able to track and even anticipate advances in wireless technology and emerging ways of implementing and using wireless services. Developing such a system in detail is beyond the scope of this committee's charge, but it is with such an objective in mind that the following items are offered as promising avenues for progress.

Avoiding the Extremes in the "Property Rights" Versus "Commons" Debate

The terms "property rights," "commons," "and greater public good" are used as shorthand for particular approaches to spectrum management. "Property rights" refers to an approach that relies on a well-specified and possibly exclusive license to operate a service using a set frequency range, location, transmitted power, and so forth. These rights can be established or transferred through an administrative proceeding, auction, or market transaction. Ideally, any of the dimensions along which the rights are defined can then be redefined through market transactions.[2] "Commons" refers to an approach that establishes a band in which those who operate devices do not need to obtain a license and instead must comply with rules that are applied to all devices operating in that channel, such as limits on transmitted power. This approach is intended to incentivize development of devices that perform better in a noisy, shared environment as an alternative to the use of market incentives for prioritizing potential sources of radiation in any given channel by the value of the communication carried in that channel.[3] "Greater public good" refers to government-sponsored free use of the spectrum for such purposes as national security, public safety, and science.

Each approach has advantages and disadvantages, transaction costs, incentives for and loci for innovation, and so forth. No one of the approaches can at present be judged to be better than the other two. Moreover, there is a much larger space of alternatives that combine attributes of these approaches, and the dividing lines between the approaches will shift as technological capabilities, deployed services, and business models evolve. These observations suggest avoiding building an overly rigid regulatory structure or relying solely on a single approach. Instead, they suggest using each approach where practically and politically feasible, and measuring and monitoring their performance, and using those results

---

[2] R.H. Coase, "The Federal Communications Commission," *Journal of Law and Economics* 2:1-40, 1959; Arthur S. De Vany, Ross D. Eckert, Charles J. Meyers, Donald J. O'Hara, and Richard C. Scott, "A Property System for Market Allocation of the Electromagnetic Spectrum: A Legal-Economic-Engineering Study," *Stanford Law Review* 21:1499, 1969; Gregory L. Rosston and Jeffrey S. Steinberg, "Using Market-Based Spectrum Policy to Promote the Public Interest," Office of Plans and Policy's Working Paper, January 1997; Gerald Faulhaber and David Farber, "Spectrum Management: Property Rights, Markets, and Commons," Office of Plans and Policy's Working Paper, 2002.

[3] Yochai Benkler, "Overcoming Agoraphobia: Building the Commons of the Digitally Networked Environment," *Harvard Journal of Law and Technology* 11(Winter):287, 1997-1998; Y. Benkler, "Some Economics of Wireless Communications," *Harvard Journal of Law and Technology* 16(1; Fall):25-83, 2002; Kevin Werbach, "Supercommons: Towards a Unified Theory of Wireless Communication," *Texas Law Review* 82:863-870, 2004.

to inform future allocations. Regulators and policy makers will need to be able to track these developments and guide where the dividing lines should be in the future.

Leveraging the Role of Standards Setting in Regulatory Decision Making

Standards are stable and well-maintained specifications that are provided by vendors, service providers, nonprofit organizations, or ad hoc organizations. They specify attributes such as interoperability and compatibility, which are often important in regulatory proceedings. Regulators often rely, at least implicitly, on technical standards that guide those building devices and services in how to comply with the regulations. This reliance on standards setting reflects two related ideas—first, that regulators may be better positioned to review technical proposals already captured in standards than to recommend specific technical approaches, and second, that it may be best to leave some of the technical details needed to define a service to standards rather than rules.

However, the process of forging consensus on standards is not easy. As in other domains, standards for wireless technologies have tended to be characterized less by engineers seeking consensus resolution of largely technical matters and more by contention among players with significant stakes in the outcome (e.g., incumbents seeking to protect their position, participants who have investments in intellectual property, or participants who have differing business interests with respect to proposed services or applications). As standards have taken on greater importance, the number of competing players and conflicting interests has grown, making the processes more cumbersome.

One risk is that the large incumbent players can dominate by virtue of their greater resources and their greater participation in standards bodies, although this risk can be partly mitigated by moving to a one company, one vote formula, but with tradeoffs. Another risk is that standards efforts could degenerate into a battle between two camps with disparate proposals that can end in a deadlock if a standard accommodates both camps—essentially a nondecision that can, at the extreme, prevent a product from coming to market or necessitate the establishment of another industry group that creates another standard that is narrow enough to be implementable.[4]

---

[4] One recent example of such a deadlock was IEEE 802.153a, an effort to establish a standard for a high-data-rate ultrawideband-based wireless personal area network. The effort ended when the standards working group was unable to decide between two competing proposals that were backed by different industry groups.

Another challenge for standards setting is that standards are most useful once a new service has already seen at least some use. That is, standards processes are most useful in helping a set of actors forge a common approach for a service that has already been developed and, at least experimentally, deployed. They are much less effective where new solutions are being sought. Related to this is the risk that some standards that are developed may never see significant adoption. Nor is the standards process always nimble or rapid; a recent example is the IEEE 802.11n standard for wireless local area networks that took many years to finally adopt, by which time interim solutions had already been widely deployed to meet the market demands for faster networking.

Understanding the Sensitivity of Innovation to Policy Decisions

The innovation process involves a number of actors, including academic researchers, small and large firms, and end users. Policy and standards setting play an important role in shaping decisions that ultimately affect innovation. Understanding the interplay between technology and policy is critical to creating effective policy. Considerations include the tension between efficiency and innovation and between the various stages of innovation. Another important consideration is that innovation depends on inputs from basic research (see below).

Ensuring Technology Expertise in the Regulatory Process

When matters requiring an evaluation of technical claims or options come before the Federal Communications Commission (FCC), the technical basis for its decisions rests on information provided in comments to the FCC and assessments made by its engineering staff (Box 3.1). The technical analyses in the submitted comments will of course tend to reflect the interests of the parties submitting those comments. The expertise of its engineering staff allows the FCC to address many specific technical issues it must grapple with regularly—for example determining the right noise figure for a particular system or the appropriate specification for adjacent channel interference.

Spectrum policy has entered an era in which many critical and strategic technical issues are likely to arise as technologies, applications, and services evolve. The FCC confronted many novel technical issues in its early days. Over time its focus has broadened and now encompasses economic and legal issues as the industries it regulates mature and broaden, issues such as broadcast media ownership and common carrier regulation. Today, qualitative and quantitative technology shifts of the sort discussed in this report and their complexity and interactions mean

> **BOX 3.1**
> **The Federal Communication Commission's**
> **Office of Engineering and Technology**
>
> The Office of Engineering and Technology (OET), the technical advisory arm of the FCC,[1] has three divisions: policy and rules, electromagnetic compatibility, and a laboratory. The policy and rules division has three branches. The spectrum policy branch covers regulations and procedures for spectrum allocation and utilization. The technical rules branch develops technical rules and standards for the operation of unlicensed RF devices. The spectrum coordination branch monitors the activities of other government agencies, particularly the National Telecommunications and Information Administration (NTIA), as well as activities of the communications industry. It is also the liaison between the FCC and the Interdepartment Radio Advisory Committee, which advises the NTIA. The electromagnetic compatibility division of OET studies radiowave propagation and communications systems characteristics; it also issues and manages experimental licenses. The laboratory division focuses mainly on testing, evaluation, and compliance. It has a technical research branch, a measurements and calibration branch, an equipment authorization branch, and a customer service branch. In 1998, OET convened a Technology Advisory Council drawn from a range of technical experts, including manufacturing, academia, communications services providers, and researchers. The council, which met regularly until July 2006, was intended to provide a means for the FCC to stay abreast of rapid advances in telecommunications technology and help inform FCC regulations in light of those advances.[2] In October 2010, a new council was appointed.
>
> ---
>
> [1] See the Office of Engineering and Technology Web site at http://www.fcc.gov/oet/.
> [2] Technology Advisory Council Charter, FCC, November 2002, available at http://www.fcc.gov/oet/tac/TACCharter_112502.pdf.

that the FCC faces new challenges of a technological nature. Examples of these complex issues that were grappled with during the work of this committee included how best to use the white space of (unused) TV channels and how best to use the 700-MHz spectrum for public safety communications.

Because it believes that the FCC would greatly benefit from enhancing its technology assessment and engineering capabilities, the committee offers several options for obtaining access to such expertise.

One option is to recruit additional top-caliber engineers and scientists to work at the FCC, perhaps for limited terms. Programs could provide early- or mid-career professionals with an opportunity to gain experience in its policy and regulatory environment or could establish rotating positions to bring in senior academic and industry experts. There is, of course, a potential for conflicts of interests to arise when staff move between government and industry, and these conflicts of interest must be carefully avoided. On balance, however, the increased flow of expertise, ideas, and perspectives seems likely to bring net benefits. The FCC has used the position of chief technologist, which has been held by several senior experts from academia and industry, as one way to bring in such expertise. The committee believes, however, that it will be necessary to create an environment that attracts more of the right talent. As things stand, for example, the committee's impression is that many in the technical community do not appear to be convinced that working at the FCC can help advance an engineering career in industry or academia.

Another option is to convene an external advisory committee that could give the FCC outside, high-level views on key technical issues. The FCC announced the appointment of a new Technology Advisory Council in October 2010, as this report was being prepared for publication.

Another option would be to add technical expertise to the staff of each commissioner. The staff members are regarded as highly competent, but most are legal professionals, not technologists. That is, although the staff members are generally knowledgeable—and often very much so—about technology, they typically do not have the advanced engineering background that may be necessary to understand and resolve complex, deeply technical issues.

Also, the FCC could tap outside technical expertise, including expertise available elsewhere in the federal government. Notably, the NTIA Institute for Telecommunication Sciences (ITS; Box 3.2) already provides considerable technical assistance to federal agencies on a cost-reimbursement basis and has done a limited amount of work for the FCC in the past. Over the years ITS has developed and maintained a strong competency in a number of technical areas related to RF communications. Strengthening the relationship between the FCC and ITS would give the FCC access to another source of independent scientific and engineering expertise on an as-needed basis. NIST, which has considerable expertise and resources for technology evaluation and is currently working in such areas as the performance of land mobile radios and their use for public safety, is another potential source of expertise. (One caveat is that the FCC's status as an independent agency rather than an executive branch agency may limit work done by the NTIA or NIST to technical and not policy matters.)

Finally, another source of outside technical expertise might be a federally funded research and development center (FFRDC). These are organizations managed by universities, industrial firms, or nonprofits and chartered to provide federal agencies with technical expertise. FFRDCs

> **BOX 3.2**
> **Institute for Telecommunications Sciences**
>
> The Institute for Telecommunication Sciences (ITS) is the research and engineering arm of the National Telecommunications and Information Administration (NTIA) in the Department of Commerce.[1] Its stated mission is to be the federal government's primary technical resource for telecommunications issues. A liaison office coordinates ITS technical research with other federal agencies. As part of its broader mission it has supported several other federal agencies, including the Departments of Defense, Homeland Security, and Transportation as well as state and local government.[2] It works through cooperative research and development agreements with the private sector (e.g., American Automobile Association, Intel, Lucent, and Motorola) and academic institutions (e.g., University of Colorado, University of Pennsylvania). ITS has also provided technical support to the FCC for specific issues such as evaluation of propagation models necessary to implement the Satellite Home Viewer Act.[3]
>
> ITS performs fundamental research and engineering with technical programs several areas directly related to wireless technology: broadband wireless, digital land mobile radio, information technology, propagation measurement and models, and spectrum research. It provides the technical resources from the United States in developing international telecommunications standards. The staff of ITS is composed mostly of scientists and engineers across a number of disciplines, including electronics engineering, math, physics, and computer science. Its stated goals reflect its engineering focus. Those goals include optimization of federal spectrum allocation methods, support for systems engineering and planning of interoperable public safety radio systems and standards (not frequency allocation, which is the purview of the FCC), improvement of network operation and management of national defense systems, and providing practical telecommunications performance measurement methods. ITS also hosts the International Symposium on Advanced Radio Technologies (ISART) conference, which annually brings together researchers, business leaders, policy makers, and regulators to discuss the future development and application of radio frequency technologies.
>
> ---
>
> [1] ITS, "FY 2007 Technical Progress Report," December 2007, available at http://www.its.bldrdoc.gov/pub/ntia-rpt/tpr/2007/07-tpr.pdf.
> [2] ITS overview brochure, available at http://www.its.bldrdoc.gov/ITS_brochure/ITS_brochure.pdf.
> [3] ITS, "Propagation Model Development," 1999, available at http://www.its.bldrdoc.gov/tpr/1999/its_e/prop_model/prop_model.html.

are able to bring in expertise on a project-by-project basis and to engage expertise that may not be available within the constraints of civil service salaries.

The committee's view is that whatever mechanisms the FCC uses to tap outside technical expertise, the goal is to strengthen capabilities for establishing appropriate high-level guidance, and not to build up an infrastructure for more detailed command-and-control regulation.

TECHNOLOGY-ENABLED POLICY OPTIONS

Considering "Open" Approaches in the Range of 20 to 100 GHz

Use is relatively sparse at frequencies of 20 to 100 GHz; commercial services in that range represent a small fraction of the services that operate below 20 GHz. The relatively high attenuation in materials—and short free space propagation in the oxygen absorption band around 60 GHz—means that propagation distances are relatively short. The ratio of antenna size to wavelength makes it practical to form very narrow beams. Together, these factors make interference inherently unlikely.

These frequencies thus represent an opportunity that stands in marked contrast to the very difficult transition problems associated with introducing new services, allocations, and sharing arrangements at lower frequencies. (Increased use of higher frequencies would, however, do little, at least in the short term, to alleviate pressures to also introduce new services at lower frequencies.) For these higher frequencies, the reduced legacy problem and lower chance for interference (in the classical sense) indicate that nontraditional ("open") approaches can predominate. Although it is an oversimplification to say this, at lower frequency the problem is dealing with the legacy, while at the higher frequency it is difficult for radios to interfere. These factors suggest that the two domains be approached differently, but the distinction has so far not been clearly articulated or incorporated into the policy-making process.

The lower bound of the range proposed for open use, 20 GHz, was selected on the basis of two factors—frequencies above 10 Ghz have only recently become practical in small devices at low cost and the region between 10 and 20 GHz is already heavily allocated, such as for Ku-band satellite transmissions between 12 and 18 GHz.

The upper bound of this range, 100 GHz, reflects what can reasonably be expected to be practical today or in the near future and the upper limit at which it is possible to have a reasonable sense of how the technology might be employed. It would thus be imprudent to recommend a particular regime for frequencies above 100 GHz, given the limited understanding of how radios might be constructed or operated in that domain, and it would be prudent to review policy in this area every few years and make adjustments as appropriate.

FCC policy has already moved toward a more flexible and adaptive approach in this frequency domain, with an unlicensed regime established at 57 to 64 GHz and licensed access to bands at 80 and 95 GHz on a first-come, first-protected basis. These measures may stimulate commercial activity and speed the deployment of new services.

At the outset, these frequencies most likely would be used for very short distances and very-high-bandwidth applications, such as in-room video distribution, because the bandwidth for gigabit and higher-rate applications is not available elsewhere. This is not to say that existing applications in those ranges would be quickly or easily replaced, but rather that over time it would be attractive to introduce new applications at 20 to 100 GHz rather than carving out the rights to introduce them at lower frequencies.

Finally, although usage at 20 to 100 GHz is relatively low compared to usage at frequencies below 20 GHz, existing users at the higher frequencies are likely to object, and some exceptions to the open rule would probably be needed to protect some existing services. For example, many satellite and military services operate in this range, mostly under NTIA jurisdiction.[5] There are also non-communications uses in this frequency range, such as radar, navigation, and other industrial, scientific, and medical uses. Recent experience in working out a sharing arrangement between WLAN and military radar use at 5 GHz suggests, in the view of the committee, both the possibilities and the potential pitfalls; an accommodation was ultimately reached but not without considerable study and delay. Because many of the existing uses above 20 GHZ are for government services, the participation of and cooperation between the NTIA and the FCC will be required to sort out the issues.

Using a Wider Set of Approaches to Mitigate Interference and a Wider Set of Parameters in Making Assignments

Interference should not be viewed simply as an overlap in frequency and space between two radios but also in terms of the ability of particular radios and radio systems to separate desired from undesired signals. Harm from interference has both technical dimensions (how well a radio or radio system can separate desired from undesired signals) and economic dimensions (the costs and distribution of costs of building,

---

[5] See, for example, Bennett Z. Kobb, *Wireless Spectrum Finder: Telecommunications, Government and Scientific Radio Frequency Allocations in the US 30 MHz-300 GHz*, McGraw-Hill, New York, 2001.

deploying, and operating a radio/radio system with particular technical characteristics that make it easier to disambiguate the signals).

Today, technology is enabling new ways of mitigating interference. The degrees of freedom available for managing interference go beyond the traditional parameters of frequency and geographical area and include amplitude, frequency, space, time, and polarization. Interference mitigation can also be thought of in terms of the behavior of radio systems rather than individual radios. In the future, coordination and cooperation are more likely to be win-win situations; a key question is how to motivate such cooperation.

Regulation is beginning to reflect these opportunities. Historically, interference between adjacent bands has been mitigated by inserting guard bands. Under recently adopted rules for the 700-MHz band, for example, there are no guard bands, leaving it up to users to figure out how to mitigate interference, whether by cooperation among users, investment in better receivers, or by other means. This is a good example of a more technology- and service-neutral approach to regulation. Rather than mandate a particular technical solution, the idea is to be flexible and allow users to find the best ways of increasing overall efficiency.

Introducing Technological Capabilities for More Sophisticated Spectrum Management

Some current and emerging technologies could make it much easier to introduce new services into crowded frequency bands. Given sufficient motivation, ingenuity, and investment, it is not possible to obtain significant improvements in communications capacity in a particular piece of spectrum, but migrating current nondigital services to digital transmission will be a major challenge, especially for specific applications like aviation radios, which have a large, politically powerful legacy base. Improvements are more feasible in bands where the disadvantages of migration are not so widely distributed across so many users, where the user base is a less potent political force, or where the market dynamics are such that end-user technology is regularly refreshed.

Smart antennas, for example, could mitigate interference problems in an overlay system. By focusing a beam from the transmitter to a receiver, devices with smart antennas can significantly reduce their overall transmission power. They could also scan their environment for other transmissions and transmit in directions that help avoid interference. These technologies are not very practical at lower frequencies but become more so at somewhat higher frequencies.

Moreover, it may be possible to incorporate more sophisticated approaches into receiver specifications established through either standards or regulations. Adaptive radios need to be able to sense their environment, negotiate with other radios, and adjust their operation accordingly. Doing so requires radios that can listen to a much wider range of signals and distinguish among various signals more accurately than is required for a conventional radio. A receiver's ability to sense a small signal in the presence of a nearby larger signal is limited both by noise, which tends to corrupt measurement of the received signal, and by the receiver's dynamic range. Thus, adaptive radios are viable only if radios meet demanding specifications for both dynamic range and noise. The problem remains of how to deal with legacy hardware, which does not have this capability built in because it was made before receiver performance was improved to exploit these opportunities.

Such higher-quality receivers also cost more, have a more complex design, and consume more power. Even small additional costs matter a great deal when service providers are fighting to save pennies. The additional investment can have a big payoff, however, if it enables new applications that are not otherwise possible.

Developing Complementary Policy to Allow Negotiation Among Users

A complement to the introduction of new technology is the creation of a policy environment in which neighbors (and others whose services experience interference) are free to negotiate a mutually acceptable outcome. This notion, first proposed by Coase,[6] provides for market negotiations to complement or replace regulatory mandates. A new arrangement may not be optimal for a given set of parties and might run the risk of becoming obsolete as technologies emerge, but such negotiations allow for flexibility in situations such as the following:

• Operator A spills over into neighbor B's spectrum in a manner that is acceptable under current regulation but is costly to neighbor B, who should be free to pay A to not spill over.
• Operator A seeks to implement a service that will interfere with operator B's service unless operator B improves the interference-rejection capabilities of its receivers. Operator A should be free to pay B for these improvements.

It is important to recognize, however, that if the transaction costs such as for bargaining are high, the bargains are likely to be less efficient. For

---
[6] R.H. Coase, "The Federal Communications Commission," *Journal of Law and Economics* 2(1):1, 1959.

example, the introduction of more sophisticated devices and network architectures could make it more difficult to know who is spilling over into a neighbor's usage rights, and who is not.

One can also envision scenarios in which such bargaining might not improve the overall efficiency of spectrum use. If license holders can negotiate with others to shut down interfering transmissions, the former will have less incentive to invest in innovative devices that can operate well in the presence of noise. Similarly, to the extent that device manufacturers know that their customers will not be able to protect themselves from interference, they will be motivated to invest in more robust, smarter devices that can give their purchasers better communications irrespective of whether or not there is an interference-reducing agreement.

Trading Absolute Outcomes for Statistically Acceptable Outcomes

Approaches that use a statistical probability of interference of less than 100 percent do not necessarily lead to ruinous outcomes that will destroy service. Rather, these approaches seek to relax constraints so as to normally (or almost always) provide good outcomes but accept poorer outcomes with acceptable probabilities and consequences. That is, the system attempts to offer optimal performance most of the time to most users and degrades softly under less optimal conditions. The difference between the approaches emphasizing absolute and acceptable outcomes regarding interference is somewhat analogous to the difference between personal auto safety (which "accepts" a certain number of accidents) and common carrier air safety (which has an explicit albeit unrealizable goal of zero accidents).

The latter approach has already been embraced in some aspects of telecommunications. The Internet's best-effort design, for example, does not guarantee quality of service yet generally provides an acceptable overall experience. Acceptance of a similar tradeoff was reflected in the market's favoring Ethernet over Token Ring technology in the early days of local area networks. Already in the wireless space, such imperfections as holes in coverage area or lower-quality audio (compared to a landline) are accepted in exchange for the convenience of mobility.

Such a relaxation of requirements could significantly open up opportunities for nonexclusive use of frequency bands. Rather than have regulators decide on acceptable quality, it might be desirable to allow licensees flexibility to negotiate mutually beneficial arrangements even though the result at times might be degraded quality.

Embracing "Design for Light" and "Design for Darkness" More Broadly in Design Concepts and Regulatory Frameworks

Many systems have been "designed for darkness"—that is, under the assumption that a particular band has been set aside for a particular service or operator and that there are no other emissions in that band. Cellular systems are a notable example of this approach. An alternative is to design for light, with an assumption of a noisy, cluttered environment. Both are reasonable design approaches for certain applications and services, but it is important to be clear about which mode is appropriate under what circumstances. The historical preference has been to design for darkness, whereas today, technological advances suggest opening up more bands in the design-for-light modality. These techniques include beam steering, enhanced signal processing, and network coordination. To design for light will require better information than is available today on sources of potential interference.

Broadening the Scope of Inquiry to Encompass Receivers and Networks of Transceivers

Much regulation has focused on transmitters, with specifications for transmission frequency and bandwidth, geographical location, and transmit power. Increasing use of new radio architectures (discussed above) suggests that the scope of inquiry be broadened to look at the properties and behaviors of receivers and networks of transceivers.

Better receiver standards would create an environment in which receiver capabilities present less of a barrier than they do today for implementing new spectrum sharing schemes. For example, it might be possible to overlay unlicensed use onto licensed use with receiver specifications written to these standards.

Expanding the scope for policy or regulation to a system of radios rather than an individual radio also would open up new opportunities. For example, a network of radios can help avoid the hidden node problem because it can use multiple network elements to listen from multiple points for transmissions. Also, a network of radios would be able to relay a transmission through hops at lower power at each node rather than directly at higher power, thus decreasing the chance it would interfere with another system. Receivers could also report on their position—for example, via embedded GPS receivers—although this capability has cost and potential privacy implications.

Exploiting Programmability

As discussed above, technology has enabled highly programmable radios. To be sure, such radios are not practical in many circumstances today because of their complexity, power use, and dollar costs, especially for mobile devices. Nonetheless, programmable radios are being used for some applications today (such as cellular base stations), and it is reasonable to expect wider use in the future. One implication of this programmability is that the radio operating parameters can be made modifiable to comply with policy or rule changes. Deployment of devices with such capabilities opens up new opportunities for more flexible regulation and for policy makers to safely work more incrementally. Namely, (1) policies would not need to be homogeneous and could be adapted to local environmental conditions such as signal density, (2) the operating rules of existing devices could be revised to accommodate new technology, and (3) devices could more easily be certified for international use because they can readily be switched to comply with local policy.

Although revisability may sound attractive, the opportunity must be weighed against some significant drawbacks. Paradoxically, rules that require revisability could actually have the effect of discouraging deployment and investment if they are seen as weakening the commitments made by regulators. The most likely scenario, if such a policy were poorly drafted, would be that most industry participants would take a wait-and-see position, which defeats the purpose of providing flexible and revisable rules for quick adoption. There are possible mechanisms to address this concern, such as offering investors compensation if the rules on which they relied are materially changed. Such mechanisms would need to be carefully considered as part of any rulemaking that sought to exploit revisability.

Exploiting Adaptive and Environment-Sensing Capabilities That Can Help Reduce Centralized Management

As agility, sensing, and coordination improve, and as etiquettes and standards for these capabilities develop, opportunities will likely arise for reduction of centralized management. Potential advantages to this approach include a lower barrier to entry (because entry either will not require engagement with a regulator for spectrum assignment or will entail negotiation with an existing license holder, or it will be easier and less costly to find an existing license holder willing to share its spectrum assignments) and flexibility of use (because operation is defined primarily by the attributes of radio equipment rather than by regulation). Potential disadvantages to this approach include uncertainty about the technical feasibility and the added costs of building more capable and robust radios. Such a shift is also predicated on resolving the issues discussed above about more robust receiver design. Some current proposals would maintain a form of centralized control but would replace regulation with much more nimble and dynamic approaches, such as services that collect and distribute information about or grant access to open channels.

Establishing Mechanisms for Dealing with Legacy Systems

In recent years, notable efforts to deal with legacy systems have included relocating microwave services to allow deployment of PCS cellular telephony and the relocation of Nextel cell services out of public safety bands. More recently, relocation of government services as well as broadcast radio services and fixed services has been undertaken for new 3G advanced wireless services bands. Having an easier process for making such changes is a critical enabler of more dynamic policies to meet changing technologies and market needs. Although there are costs and difficulties associated with relocating infrastructure elements, an even bigger legacy challenge is the need to migrate potentially millions of user-owned or user-operated devices. Among the options for dealing with legacy systems are the following:

- Commissioning independent neutral analyses to support decision making about potential interference with legacy services based on actual harm rather than political claims.
- Establishing streamlined recovery procedures. Claims of interference are inevitable where old and new systems coexist. A streamlined process would help identify, report, and resolve such claims.
- Establishing databases of legacy equipment. It is far easier to coexist with legacy systems if details about their operation are known. A lightweight system for registering systems would help to facilitate the creation of a useful database.
- Exploiting technological improvement. As radios become more capable, they will be increasingly able to coexist with existing users and services. Future policy should require or incentivize new users to coexist with existing users—for example, by making future devices more flexible (e.g., adaptable filters and oscillators and reprogrammability) so that their operation can be relocated more readily—and should avoid rules that inhibit this.

Appendixes

Appendix A

Biographies of Committee Members and Staff

David E. Liddle, *Chair*, is a general partner in the firm U.S. Venture Partners (USVP), a leading Silicon Valley venture capital firm that specializes in building companies from an early stage in digital communications, networking, wireless communications, semiconductors, technical software, and e-health. He retired in December 1999 after 8 years as CEO of Interval Research Corporation. During and after his education (B.S., electrical engineering, University of Michigan; Ph.D., computer science, University of Toledo, Ohio), Liddle has spent his professional career developing technologies for interaction and communication in research, development, management, and entrepreneurship. First, he spent 10 years at the Xerox Palo Alto Research Center and the Xerox Information Products Group, where he was responsible for the first commercial implementation of the graphical user interface and local area networking. He then founded Metaphor Computer Systems, whose technology was adopted by IBM and the company ultimately acquired by IBM in 1991. In 1992, Liddle cofounded Interval Research Corporation with Paul Allen. During his tenure, the company formed six new companies and several joint ventures based on the research conducted at Interval. He is a consulting professor of computer science at Stanford University. He has served as a director at Sybase, Broderbund Software, Metricom, Starwave, and Ticketmaster; he is currently a director with the *New York Times* and numerous early-stage companies. He was honored as a distinguished alumnus from the University of Michigan and is a member of the national advisory committee at the College of Engineering of that university. He is also a member of the advisory committee of the School of Engineering at Stanford University, and of the College of Engineering at the University of California, Berkeley. He has been elected a senior fellow of the Royal College of Art for his contributions to human–computer interaction. His current technology and investment interests are focused on signal processing, with an emphasis on wireless communications.

Yochai Benkler is the Jack N. and Lillian R. Berkman Professor of Entrepreneurial Legal Studies at Harvard Law School and faculty codirector of the Berkman Center for Internet and Society at Harvard University. His research focuses on the effects of laws that regulate information production and exchange on the distribution of control over information flows, knowledge, and culture in the digital environment. His particular focus has been on the neglected role of commons-based approaches toward the management of resources in the digitally networked environment. His books include *The Wealth of Networks: How Social Production Transforms Markets and Freedom* (2006), which received the Don K. Price Award from the American Political Science Association for best book on science, technology, and politics; the American Sociological Association's CITASA Book Award for an outstanding book related to the sociology of communications or information technology; the Donald McGannon Award for best book on social and ethical relevance in communications policy research; and was named best business book about the future by the magazine *strategy+business*. In civil society, Benkler's work was recognized by the Electronic Frontier Foundation's Pioneer Award in 2007 and by the Public Knowledge IP3 Award in 2006. Previously, Benkler was a professor at Yale University and New York University School of Law.

David Borth is an expert on wireless communications, with insight into both national security and commercial needs. He is corporate vice president and director of the Communications Research Laboratories of Motorola, Inc., a part of the company's research arm, Motorola Labs. Borth joined Motorola in 1980 as a member of the Systems Research Laboratory in corporate research and development in Schaumburg, Illinois. As a member of that organization, he has conducted research on digital modulation techniques, adaptive digital signal processing methods applied to communication systems, and personal communication systems including both cellular and PCS systems. He has contributed to Motorola's implementations of the GSM, TDMA (IS-54/IS-136), and CDMA (IS-95) digital cellular systems. In his current role, he manages a multinational (United States, Australia, France, Japan, United Kingdom) organization focusing on all aspects of communication systems ranging from theoretical systems studies to system and subsystem analysis and implementation to integrated circuit designs. Borth received his B.S., M.S., and Ph.D. degrees in electrical engineering from the University of Illinois at Urbana-Champaign. Previously, he was a member of the technical staff of the systems division of Watkins-Johnson Company and an assistant professor in the School of Electrical Engineering, Georgia Institute of Technology. Borth is a member of Motorola's Science Advisory Board Associates and has been elected a Dan Noble Fellow, Motorola's highest honorary technical award. He has been issued 31 patents and has authored or coauthored chapters of five books in addition to 25 publications. He received the Distinguished Alumnus Award from the University of Illinois Electrical and Computer Engineering Alumni Association and was elected a fellow of the Institute of Electrical and Electronics Engineers for his contributions to the design and development of wireless telecommunication systems. He is a registered professional engineer in the state of Illinois. Borth was a member of the Computer Science and Telecommunications Board from 2000 to 2003. He also served on the CSTB committee that produced the report *Information Technology for Counter-Terrorism: Immediate Action and Future Possibilities* (2003).

Robert W. Brodersen is the John R. Whinnery Distinguished Professor in the Department of Electrical Engineering and Computer Science at the University of California, Berkeley. He is also the coscientific director of the Berkeley Wireless Research Center, where he works on the application of integrated circuits as applied to personal communication systems, with an emphasis on wireless communications and low power design. Brodersen's research is focused in the areas of low-power design and wireless communications and the CAD tools necessary to support these activities. He has won best paper awards for a number of journal and conference papers in the areas of integrated circuit design, CAD, and communications, including the W.G. Baker Award in 1979. In 1982 he became a fellow of the IEEE. He was corecipient of the IEEE Morris K. Liebmann Award in 1983. He received technical achievement awards from the IEEE Circuits and Systems Society in 1986, from the Signal Processing Society in 1991, and in 1999 from the ACM Special Interest Group in Mobile Computing. Brodersen was elected a member of the National Academy of Engineering in 1988. In 1996, he received the IEEE Solid State Circuits Award. Brodersen was awarded an honorary doctorate from the University of Lund, Sweden, in 1999, and in 2000 he received the Millennium Award from the Circuits and Systems Society and the Golden Jubilee Award from the IEEE. In 2001 he was awarded the Lewis Winner Award for outstanding paper at the IEEE International Solid-State Circuits Conference. He has served on the editorial board or as a reviewer for numerous scholarly journals and publications including the IEEE *Jour-*

nal of Solid-State Circuits, IEEE Transactions on VLSI Systems, IEEE Personal Communications Magazine, and Wireless Personal Communications (Kluwer Press). He is the author or coauthor of more than 60 journal publications and 120 published conference papers and is the author, coauthor, editor, or contributor to 14 books, including An Anatomy of a Silicon Compiler (1992, Kluwer Academic Publishers) and Low Power Digital CMOS Design (1995, Kluwer Academic Publishers). He received a Ph.D. degree in engineering from the Massachusetts Institute of Technology (MIT) in 1972.

David D. Clark graduated from Swarthmore College in 1966 and received his Ph.D. from MIT in 1973. He has worked since then at the MIT Laboratory for Computer Science, where he is currently a senior research scientist in charge of the Advanced Network Architecture Group. Clark's research interests include networks, network protocols, operating systems, distributed systems, and computer and communications security. After receiving his Ph.D., he worked on the early stages of the ARPANET and on the development of token ring local area network technology. Since the mid-1970s, Clark has been involved in the development of the Internet. From 1981 to 1989, he acted as chief protocol architect in this development and chaired the Internet Activities Board. His current research area is protocols and architectures for very large, very high speed networks. Specific activities include extensions to the Internet to support real-time traffic, explicit allocation of service, pricing, and new network technologies. In the security area, Clark participated in the early development of the multilevel secure multics operating system. He developed an information security model that stresses integrity of data rather than disclosure control. Clark is a fellow of the ACM and the IEEE and is a member of the National Academy of Engineering. He received the ACM SIGCOMM Award and the IEEE Award in International Communications, as well as the IEEE Hamming Award for his work on the Internet. He is a consultant to a number of companies and serves on a number of technical advisory boards. He chaired the committee that produced the CSTB report *Computers at Risk: Safe Computing in the Information Age* and served on several committees that produced several CSTB reports.

Thomas (Ted) Darcie received his Ph.D. degree in aerospace physics from the University of Toronto in 1982. Currently, he is a professor at the University of Victoria, British Columbia, holding a Tier 1 Canada Research Chair in Optical Systems for Communications, Imaging and Sensing. Previously he worked at AT&T Bell Laboratories at Crawford Hill, Holmdel, New Jersey, where he joined the technical staff to study a wide variety of topics related to lightwave telecommunications, including fiber fabrication processes, semiconductor lasers, optical amplifiers, and numerous modulation and multiplexing techniques. He has been a lead figure in the development of lightwave systems for analog applications in cable television and wireless systems. As head of access communications research at AT&T Bell Laboratories (1989-1995), he was responsible for technology innovation in wireless, lightwave, and hybrid fiber-coax systems. He has authored more than a hundred technical publications and 25 patents spanning this broad set of technologies. From 1995 to 2002, he was vice president at AT&T Laboratories, in charge of communications infrastructure research. His research laboratory provided technology support for AT&T's diverse requirements in optical networking, broadband access, fixed wireless access, wireless LAN, and cellular systems. His team worked closely with AT&T businesses to provide technical expertise and vision and had numerous programs devoted to the evolution of mobile and broadband services, applications, and technologies. In 2002 and 2003, he was vice president for AT&T Labs Research network architecture and strategic operations planning vice president, with responsibility for connecting innovative network technologies with opportunities within AT&T's network. Darcie is an AT&T fellow and a fellow of the IEEE.

Dale N. Hatfield is an independent consultant and adjunct professor in the Department of Interdisciplinary Telecommunications at the University of Colorado at Boulder. Between December 2000 and April 2002, Hatfield served as chair of the department. Prior to joining the University of Colorado, he was the chief of the Office of Engineering and Technology at the Federal Communications Commission (FCC) and immediately before that was chief technologist at the agency. Before joining the Commission in December 1997, he was CEO of Hatfield Associates, Inc., a multidisciplinary telecommunications consulting firm in Boulder, Colorado, for 15 years. Before that, he was deputy assistant secretary of commerce for communications and information and deputy administrator of the NTIA. Before moving to the NTIA, Hatfield was chief of the Office of Plans and Policy at the FCC. In 1973 he received a Department of Commerce Silver Medal for contributions to domestic communications satellite policy and in 1999 received the Attorney General's Distinguished Service Award. In 2000, he received the Personal Communications Industry Association (PCIA) Foundation's Eugene C. Bowler Award for exceptional professionalism and dedication in government service and the FCC's Gold Medal Award for distinguished service. More recently, he received the distinguished engineer award from the University of Colorado at Boulder. He is a fellow of the Radio Club of America. In February 2001, the Federal Trade Commission appointed Hatfield as a monitor trustee for the AOL/Time Warner merger. He also serves on the board of directors of Crown Castle International and KBDI TV-12 Public Television in Denver. Hatfield holds a B.S. in electrical engineering from Case Institute of Technology and an M.S. in industrial management from Purdue University.

Michael L. Katz is the Edward J. and Mollie Arnold Professor of Business Administration of the Haas Economic Analysis and Policy Group and director of the Center for Telecommunications and Digital Convergence at the University of California, Berkeley. In 2001 and 2002, he was deputy assistant attorney general for economic analysis in the Antitrust Division of the Department of Justice. From 1994 to 1996, he was chief economist at the Federal Communications Commission. He is coeditor of the *California Management Review* and *Journal of Economics and Management Strategy*. He is a former member of the CSTB of the National Research Council. He received his Ph.D in economics from Oxford University.

Paul J. Kolodzy is currently a technology consultant in advanced wireless and networking technology, drawing on 20 years of experience in technology development for advanced communications, networking, electronic warfare, and spectrum policy for government, commercial, and academic clients. Before becoming a consultant, Kolodzy was the director of the Wireless Network Security Center (WiNSeC), a research facility at Stevens Institute of Technology that draws on wide-ranging expertise to design, develop, and evaluate technology for the secure transmission of voice, video, and data. Previously, Kolodzy had been appointed as the chair of the FCC's Spectrum Policy Task Force, which was charged with examining spectrum allocation processes and other issues so that spectrum could be put to the best use in a timely manner. Before joining the FCC, Kolodzy served as a program manager within the Advanced Technology Office of the Defense Advanced Research Projects Agency (DARPA) at the Department of Defense. At DARPA, he oversaw the initiation of next-generation communications technology, which included the neXt Generation (XG) Communications initiative. The XG project developed technology that has the potential to fundamentally change the manner in which spectrum is allocated and assigned. Kolodzy has also held positions at MIT's Lincoln Laboratory and Lockheed Martin Corporation in the development and management of advanced signal processing, RF, and EO systems. Kolodzy received a B.S. in chemical engineering from Purdue University and an M.S. and a Ph.D. in chemical engineering from Case Western Reserve University.

Larry Larson is a professor of electrical and computer engineering and director of the Center for Wireless Communications at the University of California, San Diego (UCSD). His research ranges from electronic circuits and systems to electronic devices and materials. Larson develops high-speed circuits based on InP (indium-phosphide) and GaAs (gallium arsenide) as well as silicon germanium and CMOS technology. He also explores applications for micromachining technology in the manufacture of high-speed integrated circuits and studies new packaging technology for them. Larson's current research is specifically focused on low-power circuit design and RF design techniques for wireless communications. He recently completed *CDMA Mobile Radio Design*, a book on how to design the hardware and software for wireless handsets based on code-division multiple access technology. CDMA is the foundation of all 3G wireless technologies, including Europe's W-CDMA standard and CDMA2000. As director of the industry-sponsored Center for Wireless Communications (CWC) at UCSD, he oversees a wide range of ongoing research projects, with funding from CWC's 17 corporate members. He is the first holder of the communications-industry-endowed chair at the Jacobs School. He joined the UCSD faculty in 1996 after a 16-year career at Hughes Research Laboratories. There, he pioneered the development of analog integrated circuits and low-noise high-electron-mobility transistors in III-V technology, as well as microwave integrated circuits in SiGe HBT technology and RF MEMS technology. Larson received a Ph.D. from the University of California, Los Angeles, in 1986. He is an IEEE fellow and co-winner of the 1996 Hughes Electronics Lawrence Hyland Patent Award and the 1999 IBM Microelectronics Excellence Award.

David P. Reed is a senior vice president in the chief scientist group at SAP Labs and an adjunct professor at the MIT Media Laboratory. He was previously a fellow at HP Labs. Reed's work focuses on using digital technology to transform the design of technological, business, and social systems. His explorations center on exploiting new information technologies that enable people to be more effective, including mobile computing; highly scalable wireless networking; group information sharing; pervasive networking; video media processing; and infrastructures for electronic commerce. Reed spent 4 years at Interval Research Corporation exploring portable and consumer media technology. For 7 years before joining Interval, Reed was vice president and chief scientist for Lotus Development Corporation, where he led the design and implementation of key products, including 1-2-3, and technical business strategy. Reed was also a professor in MIT's Laboratory for Computer Science. He is coinventor of the end-to-end argument, often called the fundamental architectural principle of the Internet. He holds a B.S. in electrical engineering and M.S. and Ph.D. degrees in computer science and engineering from MIT.

Gregory Rosston is the deputy director of the Stanford Institute for Economic Policy Research. His research focuses on industrial organization, antitrust, and regulation. He has written numerous articles on competition in local telecommunications, implementation of the Telecommunications Act of 1996, and auctions and spectrum policy. He has also co-edited two books, including *Interconnection and the Internet: Selected Papers from the 1996 Telecommunications Policy Research Conference*. Before joining Stanford University, Rosston served as deputy chief economist of the FCC, where he helped to implement the Telecommunications Act. In this work, he helped to design and write the rules that the FCC adopted as a framework to encourage efficient competition in telecommunications markets. He also helped with the design and implementation of the FCC's spectrum auctions. Rosston received his Ph.D. in economics from Stanford University and his A.B. in economics with honors from the University of California, Berkeley.

David Skellern is CEO of National ICT Austrialia (NICTA). Skellern began his career in 1974 at the University of Sydney, where he spent a decade designing, building, and commissioning instrumentation and extensions for the Fleurs Synthesis Radiotelescope, one of Australia's pioneering giant radiotelescopes. From 1983 to 1989 he held various academic appointments as a staff member of Sydney University's Electrical Engineering Department. In 1989, Skellern took up the chair of electronics at Macquarie University. He also spent considerable time working in industry as a visiting researcher, including more than 2 years at Hewlett-Packard Laboratories. In 1997 he cofounded the Radiata group of companies in Australia and the United States, established to commercialize the results of the WLAN research project that he led at Macquarie University in collaboration with the Commonwealth Scientific and Industrial Research Organization. Over the next 3 years he played an integral role in building a successful company with a team of 65. In September 2000 Radiata demonstrated the world's first chip-set implementation of the 54 Mbps IEEE 802.11a, a high-speed WLAN standard. Radiata was acquired by Cisco Systems, Inc., in 2001 for 565 million Australian dollars, at which time Skellern joined Cisco and subsequently moved to the United States as technology director of the Wireless Networking Business Unit. Skellern was appointed to the NICTA board in 2003. He received a B.Sc. (computer science and mathematics) in 1972, a B.E. (electrical engineering) in 1974, and a Ph.D. in 1985 from the University of Sydney.

Staff

Jon Eisenberg is director of the Computer Science and Telecommunications Board of the National Academies. At CSTB, he has been the study director for a diverse body of work, including a series of studies exploring Internet and broadband policy and networking and communications technologies, and a study of how to use information technologies to enhance disaster management. From 1995 to 1997 he was a AAAS Science, Engineering, and Diplomacy Fellow for the U.S. Agency for International Development, where he worked on environmental management, technology transfer, and information and telecommunications policy issues. He received his Ph.D. in physics from the University of Washington in 1996 and a B.S. in physics with honors from the University of Massachusetts at Amherst in 1988.

Appendix B

Speakers at Meetings

Although the briefers and workshop speakers listed below provided much useful information of various kinds to the committee, they were not asked to endorse the report's conclusions or recommendations, nor did they see the final draft of this report before its release.

October 23-24, 2003
The National Academies
Washington, D.C.

Joseph B. Evans, National Science Foundation
Michael D. Gallagher, National Telecommunication and Information
    Administration
Paul Kolodzy, Wireless Network Security Center, Stevens Institute of
    Technology
James A. Lewis, Center for Strategic and International Studies
James H. Snider, New America Foundation
Peter Tenhula, Federal Communications Commission

January 29-30, 2004
Stanford Universitys
Palo Alto, California

Bob Brodersen, University of California, Berkeley
Michael Howse, PacketHop
Devabhaktuni Srikrishna, Tropos Networks

February 12-13, 2004
Workshop
The National Academies
Washington, D.C.

Siavash Alamouti, Vivato
Richard Barth, Department of Commerce
Samuel W. Bodman, Department of Commerce
David G. Boyd, SAFECOM Program, Department of Homeland Security
Thera Bradshaw, City of Los Angeles Information Technology Agency
Charles N. Brownstein, Computer Science and Telecommunications Board
Duane Buddrius, Alvarion, Inc.
Jim Bugel, Cingular Wireless LLC
Leigh Chinitz, Proxim Corporation
Mark Cooper, Consumer Federation of America
Diane Cornell, Cellular Telecommunications & Internet Association
Thomas Cowper, Statewide Wireless Network, New York's Office of Technology
David Donovan, Association for Maximum Service Television, Inc.
Tyler Duvall, Department of Transportation
Harold Feld, Media Access Project
Bruce Fette, General Dynamics Decision Systems
Michael Gallagher, National Telecommunications and Information Administration
Merri Jo Gamble, Department of Justice
Michael Green, Atheros Communications
Kalpak Gude, PanAmSat
Robert Gurss, Association of Public-Safety Communications Officials International
Dewayne Hendricks, Dandin Group, Inc.
Bradley Holmes, Arraycomm, Inc.
Nancy Jesuale, Net City Engineering, Inc.
Kevin Kahn, Intel
Julius Knapp, Federal Communications Commission
Robert LeGrande, Office of the Chief Technology Officer, District of Columbia
Pat Mahoney, Iridium Satellite LLC
Preston F. Marshall, Advanced Technology Office, Defense Advanced Research Projects Agency
William Moroney, United Telecom Council and United Power Line Council
John Muleta, Wireless Telecommunications Bureau, Federal Communications Commission

Glen Nash, Department of General Services, State of California
Scott Pace, National Aeronautics and Space Administration
Carl Panasik, Texas Instruments
Andrea Petro, Office of Management and Budget
Marilyn Praisner, Montgomery County Council
Dipankar Raychaudhuri, Wireless Information Network Lab, Rutgers University
Paul Rinaldo, American Radio Relay League
George (Gee) Rittenhouse, Lucent Technologies
Kenneth Ryan, Comsearch
Greg Schmidt, LIN Television Corporation
David Siddall, Paul, Hastings, Janofsky, and Walker, LLP
Jim Smoak, Verizon Wireless
Carl Stevenson, Agere Systems
Karen St. Germain, National Oceanic and Atmospheric Administration
Thomas Walsh, Boeing Space and Communication Spectrum Management
Jennifer Warren, Lockheed Martin
Charles Wheatley, Qualcomm
Donald Willis, Federal Aviation Administration
Moe Z. Win, Massachusetts Institute of Technology
Badri Younes, Department of Defense

July 22-23, 2004
University of California, San Diego
San Diego, California

Bob Brodersen, University of California, Berkeley
Michael Chartier, Intel
Robert Matheson, National Telecommunications and Information Administration
Allen Petrin, Georgia Institute of Technology
Chuck Wheatley, Qualcomm

Appendix C

Statement of Task

An expert committee will be convened to conduct a comprehensive assessment of wireless technology and application trends and their implications for spectrum management and policy. The study will be grounded in an assessment of how technology capabilities are evolving, including the implications of emerging technologies (such as software radios, smart antennas, and other intelligent signal processing), architectural alternatives (such as base station-based and peer-to-peer), services (such as 3rd and 4th generation mobile, local area networking, and fixed broadband), and applications. Building on this technology assessment, the study will also examine the interplay between the technical, economic, and policy issues. Key policy issues to be considered include spectrum supply and demand, alternative spectrum management approaches (including unlicensed approaches), standards-setting processes and forums, and how the international environment is evolving and affecting U.S. policy options. The committee is seeking broad input on these issues from academic and industry experts and diverse stakeholders.

In addition, the committee will convene a workshop examining the present and prospective needs of public- and private-sector spectrum users and technology and policy options for more efficient and effective use of spectrum. The scope of topics to be considered in the workshop will be very similar to that of the broader study, but with additional emphasis on federal, state, and local government spectrum uses.

The committee will issue a brief report of the workshop. The committee will also produce a final report with consensus findings and recommendations.